SpringerBriefs in Earth System Sciences

South America and the Southern Hemisphere

Series editors

Gerrit Lohmann, Bremen, Germany
Lawrence A. Mysak, Montreal, Canada
Justus Notholt, Bremen, Germany
Jorge Rabassa, Ushuaia, Argentina
Vikram Unnithan, Bremen, Germany

More information about this series at http://www.springer.com/series/10032

Teresa Radziejewska

Meiobenthos in the Sub-equatorial Pacific Abyss

A Proxy in Anthropogenic Impact Evaluation

 Springer

Teresa Radziejewska
Palaeoceanology Unit
  Faculty of Geosciences
University of Szczecin
Szczecin
Poland

ISSN 2191-589X   ISSN 2191-5903   (electronic)
ISBN 978-3-642-41457-2 ISBN 978-3-642-41458-9   (eBook)
DOI 10.1007/978-3-642-41458-9

Library of Congress Control Number: 2014943740

Springer Heidelberg New York Dordrecht London

Springer is part of Springer Science+Business Media (www.springer.com)

To my Mother

Acknowledgments

It is my pleasant duty to express sincere gratitude to all those whose good will, encouragement, support, inspiration, assistance, endurance, and readiness to share information and knowledge has made preparation of this work possible and worthwhile. The long list of those individuals begins with Prof. Ryszard Kotliński, formerly the Interoceanmetal Joint Organization (IOM) Director-General whom I owe sincere thanks for the support, for sharing knowledge, and for placing a lot of material at my disposal to be used in this book. The list includes my former IOM colleagues and cruise companions: Drs. Valcana Stoyanova, Igor Modlitba, Antonin Pařizek, Jan Horniš, and Georgui G. Tkatchenko; the crew members and scientific teams on board RVs *Yuzhmorgeologiya* and *Professor Logachev*, Mr. Vyacheslav Melnik in particular, and Messrs Bret Ray (Sound Ocean Systems Inc.), Huang Yongyang, Chen Yongchin, and Wang Chungsheng. I am grateful to Prof. Idzi Drzycimski for the encouragement and for reading and commenting on earlier drafts of some of the chapters. Ms. Halina Dworczak and Dr. Joanna Rokicka-Praxmajer of the West Pomeranian University of Technology in Szczecin, Poland, and Dr. Maria Szymelfenig of the Institute of Oceanography, University of Gdańsk, Poland are thanked for their assistance at various stages of the work. I am greatly indebted to Dr. Valentina V. Galtsova (State University of Hydrography, St. Petersburg, Russia) for her skilful taxonomic work, insights and friendship; the taxonomic assistance of Dr. Lena V. Kulangieva (Zoological Institute, Russian Academy of Sciences, St. Petersburg, Russia) is gratefully acknowledged as well. Many thanks to Dr. Thomas Soltwedel of the Alfred Wegener Institute—Helmholtz Centre for Polar and Marine Research (AWI) in Bremerhaven (Germany) and to AWI librarians for facilitating my literature search. I value greatly discussions with and feedback from Dr. Erdogan Ozturgut (formely of NOAA) as well as Profs. Hjalmar Thiel (formerly of AWI) and David Thistle of the Florida State University in Tallahassee, Florida, USA, and Mr. Tomohiko Fukushima of the Marine Minerals Agency of Japan (MMAJ). My thanks are also due to my colleagues at the Institute of Marine and Coastal Sciences, University of Szczecin, Szczecin, Poland, particularly to Dr. Brygida Wawrzyniak-Wydrowska who helped in more ways than one, as well as to

Ms. Aleksandra Kaniak and Messrs Tomasz Zawiślak and Łukasz Cieszyński for all their technical support and invaluable contributions to the graphical aspect of this book. I thank the Interoceanmetal Joint Organization for placing their materials at my disposal. I wish to acknowledge the support provided by the University of Szczecin (Palaeoceanology Unit) statutory funds for research. I would also like to thank Dr. Johanna Schwarz, my Springer editor, for her patience. Last but not least, I owe immense thanks to Prof. Izabella Dunin-Kwinta, formerly of the Maritime University, Szczecin, Poland for steering me into the deep sea, and to my husband, Stefan Matalewski, for always standing by.

Contents

Prologue

The deep seafloor (understood here as the oceanic bottom at depths exceeding 1,000 m) is the largest habitat on Earth. It is a highly diverse part of the oceanic mega-system and comprises a number of sub-habitats, each supporting specific benthic communities (e.g. Buhl-Mortensen et al. 2010; Vanreusel et al. 2010). These communities and their spatial variability are assumed to be, to a great extent, shaped by the type of their sedimentary environment, primarily the bottom sediment and its properties (Gray and Elliott 2009; Greene et al. 1999). Sediment properties in turn are a net result of numerous and complex interactions involving, i.e. the depth gradient, sedimentation regime, proximity to land masses, near-bottom hydrodynamics, underwater tectonics, seismicity, and volcanism (Brown et al. 1989; Seibold and Berger 1993). Current knowledge on these processes and on the way they affect the deep seafloor environments and the life forms these support is, despite extensive research efforts, particularly in the recent decades, still incomplete (Ramirez-Llodra et al. 2010). However, in the currently expounded framework of appreciation for ocean goods and services (e.g. Armstrong et al. 2012), there is a growing realisation worldwide that the deep seafloor can, and should, fulfill an important role—that of a provider of material goods in the form of biotic and abiotic resources (Armstrong et al. 2012).

With respect to the latter, it is already well established that the deep seafloor is a repository of a variety of mineral resources some of which are—or have a potential to be—of a key importance for human activities on Earth, both at present and in the future (Gage and Tyler 1991; Glover and Smith 2003; Summerhayes 1996). In addition to the raw materials such as placer deposits, oil, and gas that are being exploited at present, the seafloor stores certain resources which—when their terrestrial equivalents will have been exhausted and analogues will be impossible to find or will be too costly to develop—are an important wealth to be tapped (Glover and Smith 2003). This type of resources includes oceanic polymetallic nodules, also known as (ferro)manganese nodules (Hein and Petersen 2013; Hoffert 2008; Kotliński 1998a). Discovered during the 1882–1876 "Challenger" expedition (Hoffert 2008), the nodules occur as more or less spherical concretions (Fig. 1)

Fig. 1 *Left* Polymetallic nodules from the sub-equatorial NE Pacific's Clarion-Clipperton Fracture Zone (CCFZ) on the ship's deck, assembled for analysis (*Photo* T. Radziejewska); *Right* a close-up of a polymetallic nodule, with abundant epifauna (*Photo* B. Wawrzyniak-Wydrowska)

Fig. 2 A fragment of nodule-bearing seafloor in the CCFZ, featuring a holothurian (megabenthos) (*Photo* courtesy of IOM)

consisting of metal hydroxides, ores, and a variety of other minerals (Hein and Petersen 2013) strewn on the oceanic bottom (Fig. 2), mostly in the oceanic abyss.

Particularly abundant nodule deposits have been discovered on the deep seafloor of some areas in the Atlantic, Indian, and Pacific Oceans, characterised by low sedimentation rates (Hoffert 2008; Kotliński 1998a, 1999; Ross 1980; Seibold and Berger 1993). The most abundant nodule deposits in the Pacific Ocean are found on the abyssal bottoms of the Peru Basin in the southern part and between the Clarion and Clipperton fractures (the Clarion-Clipperton Fracture Zone, CCFZ) in NE Pacific (Hein and Petersen, 2013; Kotliński 1999; Glover and Smith 2003).

It is now commonly accepted that commercial mining of polymetallic nodules is imminent, although the commencement of mining operations still remains a matter of the future (Glover and Smith 2003; Berge et al. 1991; Hoffert 2008; Kotliński 1998b; Padan 1990; Thiel et al. 1992, 1998; Sharma 2011). In view of the imminence of mining, it has been deemed necessary to establish a legal framework for this activity (Hoffert 2008). Such framework has been indeed provided by the United Nations Convention on the Law of the Sea (UNCLOS) and regulations and activities stemming from it. Based on the UNCLOS provisions, the International Seabed Authority (ISA) was set up as an international body charged with management of the seafloor resources beyond the national jurisdiction of coastal states; those resources are recognised as the common heritage of Mankind (Hoffert 2008; Kotliński 1998b, 1999). The ISA's mission includes, i.a. making sure that the heritage is managed in a responsible and sustainable manner (http://www.isa.org.jm).

Responsible approach to developing seafloor resources, including the polymetallic nodules, requires that potential effects, particularly the adverse ones, of such an activity for the bottom habitat and its communities be realised, assessed and minimised (Berge et al. 1991; Jumars 1981; National Research Council 1984; Ozturgut et al. 1978; Thiel 1992; Thiel et al. 1992, 1998). At the present stage of preparations to commercial mining, the major actors (formerly termed the "pioneer investors" and currently known as the ISA contractors) are aware that they will be obliged to assess environmental consequences of mining, that is changes to the abiotic environment (bottom sediment and water column) and to the benthic and pelagic biota, resulting from mining activities. This obligation has been formalised in specific guidelines required by UNCLOS provisions and prepared by ISA, constituting the Mining Code (ISBA 2013) as well as in the guidelines pertaining to the assessment of environmental consequences of exploration activities in nodule-bearing areas (ISBA 2010). Efforts towards establishing a regulatory legal framework for the development of polymetallic nodule deposits have been recently summarised in ISA Technical Study No. 11 (ISA 2013).

As a step towards fulfilment of their legally binding obligations, some states and consortia enjoying the prospect of future commercial activities in their respective claim areas in the abyssal nodule-bearing oceanic bottom (Fig. 3) had embarked on research programmes aimed at collecting information on natural variability of the abyssal environment and on potential effects of mining-related intervention into it (Kotliński and Stoyanova 1998; Schriever et al. 1997; Sharma et al. 2001; Thiel 2001; Yamazaki and Kajitani 1999; Zhou 1997).

It was during one of such programmes, conducted by the Interoceanmetal Joint Organization (IOM)—an intergovernmental body (an ISA contractor) set up in 1985 to prepare commercial mining of polymetallic nodules in an area located within the sub-equatorial NE Pacific's Clarion-Clipperton nodule field (Kotliński 1998a, b)—that the present author was, in 1995–2000, a member of a team studying deep-sea environment and its biota (Radziejewska 2002; Radziejewska and Kotliński 2002; Radziejewska and Stoyanova 2000). The studies were aimed at collecting data that would augment the then existing body of knowledge on the

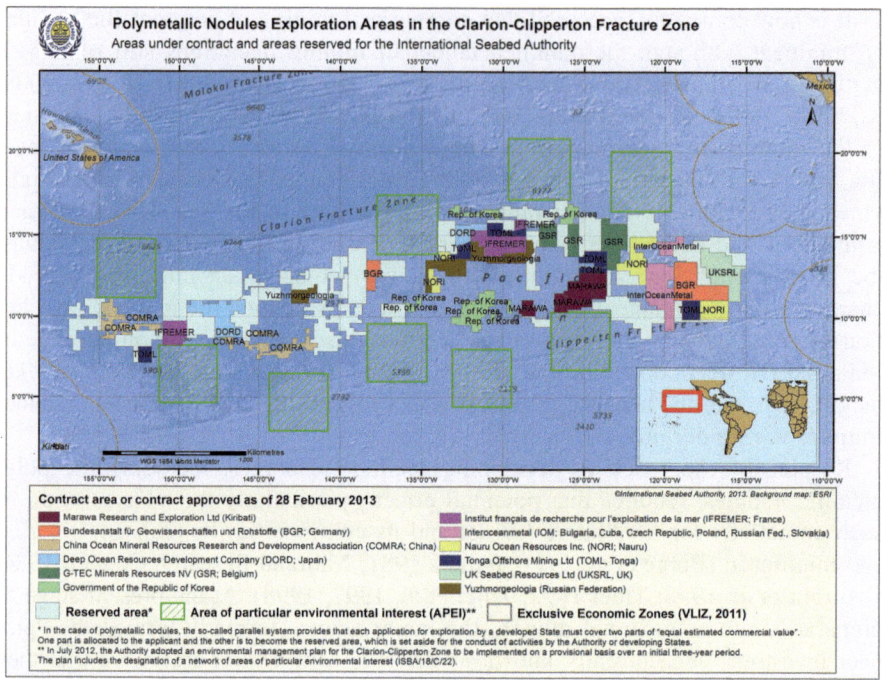

Fig. 3 Subdivision of the CCFZ (www.isa.org.jm)

Pacific abyssal environment and its communities, and at providing insights into the structure and responses of certain compartments of benthic faunal assemblages to man-made disturbance. It was hoped that, in addition to its inherent cognitive value, this information should aid in strengthening the groundwork on which to base future monitoring and assessment activities focused at evaluating consequences of polymetallic nodule mining-related environmental disturbance.

The first several years of the twenty-first century have been, and are, witnessing a renewed international interest in securing rights and claims to the oceanic nodule deposits, and an increasingly higher number of states and consortia rush to join the group of ISA contractors (e.g. Lodge et al. 2014; Schrope 2013). At the same time, fortunately, there is a growing realisation that the habitat where such activities are going to be pursued, and its communities, are—although vastly understudied— very fragile and irreplaceable. This realisation has sparked interest in, and resulted in pleas for, studies and conservation plans and concepts with regard to the deep-sea benthic habitats and communities in general (Mengerink et al. 2014; Van Dover 2011; Van Dover et al. 2014) and to those specific for the nodule-bearing areas in particular (Wedding et al. 2013). As elsewhere in such studies, particularly when the assessment of the magnitude of impact likely to be produced is to be a result, appropriate indicators, or proxies, will have to be selected, tested, and used (Jørgensen et al. 2010). A suite of such proxies, abiotic and biotic, has been in use in coastal areas, and includes variables related to the structure of benthic communities, including the meiobenthos. As the meiobenthos is a likely candidate for a proxy also in the deep-sea impact assessments, a synthesis of past research involving this category of benthos in nodule-bearing areas as a response variable in the assessment of effects produced by seafloor disturbance resulting from mining-like activities—the purpose and focus of this book—seems justified and timely.

References

Armstrong C, Foley N, Tinch R et al (2012) Services from the deep. Steps towards valuation of deep sea goods and services. Ecol Serv 2:2–13

Berge S, Markussen JM, Vigerust G (1991) Environmental consequences of deep seabed mining. Problem areas and regulations. Fridtjof Nansen Institute, Lysaker

Brown J, Colling A, Park D et al (1989) Ocean chemistry and deep sea sediments. The Open University, Milton Keynes

Buhl-Mortensen L, Vanreusel A, Gooday AJ et al (2010) Biological structures as a source of habitat heterogeneity and biodiversity on the deep ocean margins. Mar Ecol 31:21–50

Gage JD, Tyler PA (1991) Deep-sea biology: a natural history of organisms at the deep-sea floor. Cambridge University Press, Cambridge

Glover AG, Smith CR (2003) The deep-sea floor ecosystem: current status and prospects of anthropogenic change by the year 2025. Environ Conserv 30:219–241

Gray JS, Elliott M (2009) Ecology of marine sediments. From science to management. 2nd edn. Oxford University Press, Oxford

Greene HG, Yoklovich MM, Starr RM et al (1999) A classification scheme for deep seafloor habitats. Oceanol Acta 22:663–678

Hein JR, Petersen S (2013) The geology of manganese nodules. In: Baker E, Beaudoin Y (eds) Deep sea minerals: manganese nodules, a physical, biological, environmental, and technical review, vol 1B. Secretariat of the Pacific Community, GRID-Arendal

Hoffert M (2008) Les nodules polymétalliques dans les grands fonds océaniques. Une extraordinaire aventure minière et scientifique sous-marine. Société Géologique de France, Vuibert

ISA (2013) Towards the development of a regulatory framework for polymetallic nodule exploitation in the area. ISA Technical Study 11, International Seabed Authority, Kingston, Jamaica

ISBA (2010) Recommendations for the guidance of contractors for the assessment of the possible environmental impacts arising from exploration for polymetallic nodules in the area. ISBA/16/LTC/7. https://www.isa.org.jm/files/documents. Accessed 3 Apr 2014

ISBA (2013) Regulations on prospecting and exploration for polymetallic nodules in the area. ISBA/19/C/17. https://www.isa.org.jm/files/documents. Accessed 3 Apr 2014

Jørgensen SE, Xu F-L, Marques JC et al (2010) Application of indicators for the assessment of ecosystem health. In: Jørgensen SE, Xu F-L, Costanza R (eds), Handbook of ecological indicators for assessment of ecosystem health, 2nd edn. CRC Press, Boca Raton

Jumars PA (1981) Limits to predicting and detecting benthic community responses to manganese nodule mining. Mar Mining 3:213–229

Kotliński R (1998a) Konkrecje polimetaliczne. In: Depowski S, Kotliński R, Rühle E et al (eds) Surowce mineralne mórz i oceanów. Wydawnictwo Naukowe Scholar, Warszawa

Kotliński R (1998b) The present state of knowledge on oceanic polymetallic ores as exemplified by Interoceanmetal Joint Organization's activity. Mineral Pol 29:77–89

Kotliński R (1999) Metallogenesis of the world's ocean against the background of oceanic crust evolution. Pol Geol Inst Spec Pap 4.

Kotliński R, Stoyanova V (1998) Physical, chemical, and geological changes of marine environment caused by the benthic impact experiment at the IOM BIE site. In: Chung JS, Olagnon M, Kim CH et al (eds), Proceedings of 8th ISOPE Conference, vol 2. Montreal, Canada, pp 277–281

Lodge M, Johnson D, Le Gurun G et al (2014) Seabed mining: International Seabed Authority environmental management plan for the Clarion-Clipperton Zone. A partnership approach. Mar Policy 49:66–72

Mengerink K, Van Dover CL, Ardron J et al (2014) A call for deep-ocean stewardship. Science 344:696–698

National Research Council (1984) Deep seabed stable reference areas. National Academy Press, Washington, D.C.

Ozturgut E, Anderson GD, Burns RE et al (1978) Deep ocean mining of manganese nodules in the North Pacific: pre-mining environmental conditions and anticipated mining effects. NOAA Techn Mem, ERL MESA-33

Padan JW (1990) Commercial recovery of deep seabed manganese nodules: twenty years of accomplishments. Mar Mining 9:87–103

Radziejewska T (2002) Responses of deep-sea meiobenthic communities to sediment disturbance simulating effects of polymetallic nodule mining. Int Rev Hydrobiol 87:459–479

Radziejewska T, Kotliński R (2002) Acquiring marine life data while experimentally assessing environmental impact of simulated mining in the deep sea. ICES, CM 2002/L:01

Radziejewska T, Stoyanova V (2000) Abyssal epibenthic megafauna of the Clarion-Clipperton area (NE Pacific): changes in time and space versus anthropogenic environmental disturbance. Oceanol Stud 29:83–101

Ramirez-Llodra E, Brandt A, Danovaro R et al (2010) Deep, diverse and definitely different: unique attributes of the world's largest ecosystem. Biogeosci 7:2851–2899

Ross DA (1980) Opportunities and uses of the ocean. Springer, New York

Schriever G, Ahner A, Bluhm H et al (1997) Results of the large-scale deep-sea experimental study DISCOL during eight years of investigation. In: Chung JS, Das BM, Matsui T et al (eds), Proceedings of 7th (2006) ISOPE, vol 1. Honolulu, Hawaii, pp 438–444

Schrope M (2013) UK company pursued deep-sea bonanza. Nature 496:294

Seibold E, Berger WH (1993) The sea floor. An introduction to marine geology, 2nd edn. Springer, Berlin

Sharma R (2011) Deep-sea mining: economic, technical, technological, and environmental considerations for sustainable development. Mar Technol Soc J 45:28–41

Sharma R, Nath BN, Parthiban G et al (2001) Sediment redistribution during simulated benthic disturbance and its implications on deep seabed mining. Deep-Sea Res II 48:3363–3380

Summerhayes CP (1996) Ocean resources. In: Summerhayes CP, Thorpe SA (eds) Oceanography. An illustrated guide. Manson Publishing, London

Thiel H (1992) Deep-sea environmental disturbance and recovery potential. Int Rev ges Hydrobiol 77:331–339

Thiel H (2001) Use and protection of the deep sea—an introduction. Deep-Sea Res II 48:3427–3431

Thiel H, Foell EJ, Schriever G (1992) Potential environmental effects of deep seabed mining. University of Hamburg, Hamburg

Thiel H, Angel MV, Foell EJ et al (1998) Environmental risks from large-scale ecological research in the deep-sea: a desk study. Office Off Publ Europ Comm, Luxembourg

Van Dover CL (2011) Mining seafloor massive sulphides and biodiversity: what is at risk? ICES J Mar Sci 68:341–348

Van Dover CL, Aronson J, Pendleton L et al (2014) Ecological restoration in the deep sea: Desiderata. Mar Policy 44:98–106

Vanreusel A, Fonseca G, Danovaro R et al (2010) The contribution of deep-sea macrohabitat heterogeneity to global nematode diversity. Mar Ecol 31: 6–20

Wedding LM, Friedlander AM, Kittinger JN et al (2013) From principles to practice: a spatial approach to systematic conservation planning in the deep sea. Proc Roy Soc Ser B 280: 20131684. doi.org/10.1098/rspb.2013.1684

Yamazaki T, Kajitani J (1999) Deep-sea environment and impact experiment to it. In: Chung JS, Matsui T, Koterayama W (eds), Proceedings of 9th ISOPE Conference, vol 1. Brest, France, pp 374–381

Zhou H-Y (1997) Chinese environmental investigations related to potential impact of deep-sea mining. In: Chung JS, Das BM, Matsui T et al (eds) Proceedings of 7th ISOPE Conference, vol 1. Honolulu, Hawaii, pp 496–498

Chapter 1
Introduction

Abstract The meiobenthos (also called the benthic meiofauna) is a heterogenous group of benthic organisms, both protists and metazoans. Initially distinguished among the benthic organisms on account of their size [as organisms retained on 0.063 (0.032)–1.00 (0.500 or 0.250) mm mesh size sieves], the grouping has become recognised as a distinct ecological category, important by its major contribution to benthic metabolism and secondary production. While marine ecological research usually addresses entire meiobenthic communities considered as assemblages of interacting components represented by high-rank taxonomic units called the major taxa (phyla, orders, families), there is a general awareness of an immense taxonomic richness (diversity) those taxa represent. Whenever detailed taxonomic studies on the meiobenthos have been carried out, a great number of new species, genera and higher-rank taxa has been described. However, the knowledge of this diversity, particularly in the deep sea, is still greatly limited. Ecological research on the meiobenthos revealed the grouping to be a sensitive indicator of environmental changes. Consequently, the meiobenthos is being increasingly frequently used in monitoring and evaluating impacts of factors that disturb the natural state of sedimentary environment. The reliability of such evaluations may be enhanced by refining the resolution of taxonomic analyses and by coupling them with information on functional traits of the meiobenthic taxa present in an assemblage. While such approach is gaining popularity in research on coastal areas, it is still very rare in the deep sea, although the meiobenthos-related variables have been used in evaluating impacts in the deep sea.

Keywords Meiobenthos · Diversity · Functional diversity · Taxonomic sufficiency · Disturbance indication

© The Author(s) 2014   1
T. Radziejewska, *Meiobenthos in the Sub-equatorial Pacific Abyss*,
SpringerBriefs in Earth System Sciences, DOI 10.1007/978-3-642-41458-9_1

The meiobenthos (also called the benthic meiofauna) is a term denoting a very heterogenous group of benthic organisms, both protists and metazoans. The metazoans represent an array of phyla (e.g. the Nematoda, Gnathostomulida, Gastrotricha, Kinorhyncha, Loricifera, Tardigrada) and lower taxonomic units (classes, orders, families), collectively termed the "major taxa" (Giere 2009). In addition, members of certain taxa are assigned to the meiobenthos only when young (and small); these are termed the temporary, as opposed to permanent, meiobenthos (McIntyre 1969). The meiobenthos was initially distinguished among other seafloor organisms on account of the size. The meiobenthic size is usually defined by the range of mesh sizes recommended for use when separating organisms from the sediment; this range spans, overall, from 0.032 to 1 mm (Giere 2009; McIntyre 1969). As the deep-sea benthic organisms tend to be generally smaller ("miniaturised") compared with those dwelling in shallower areas, Thiel (1983) proposed to modify the deep-sea meiobenthic size range and to use sieves of 0.030–0.500 mm mesh sizes to retain sediment from which the meiobenthic organisms are extracted.

The body size has been, and still is, the basic criterion with which to set off the meiobenthos from the remaining benthic fauna (Mare 1942). However, as the study of meiobenthic assemblages was progressing, it turned out that the body size is by no means the only criterion. The present knowledge of biology and ecology of the meiofauna, although greatly inadequate in view of the actual needs and the number of question waiting to be answered, allows to treat the group as an ecological entity and to conclude that the meiobenthos have a number of characteristics different from those of other benthic animals. The grouping is distinct among other benthos on account of its taxonomic composition (numerous taxa, e.g. harpacticoid copepods, the Kinorhyncha, the Loricifera never exceed the meiobenthic size), life cycle characteristics (e.g. development confined to the sedimentary environment—the absence of pelagic larvae), evolution, and abundance (frequently about 2–3 and more orders of magnitude higher than that of larger animals, the macrobenthos). Conclusions drawn from studies on the size structure of benthic communities (Drgas et al. 1998; Duplisea and Drgas 1999; Schwinghamer 1981; Warwick 1984; Warwick et al. 1986), in which the meiobenthos stands out as a distinct part of the benthos size spectrum, too, allow to regard the bottom meiofauna as a well-defined ecological (at least body size-based) category in aquatic ecosystems (but see Bett 2013 who advised caution in dividing the integrated ranges of benthic metazoan body size spectrum into distinct entities when too few sieves are used for physical separation of size classes).

While accepting the meiobenthos as an ecological category within the benthos, various authors attempted to determine the role of meiofauna in partitioning material and energetic resources of the sediment, and to assess those organisms' contribution to energy flow in benthic communities. Some authors (e.g. Drgas 1999; Duplisea and Hargrave 1996) argue that this role is considerable and commensurate with the contribution of the meiobenthos to the total benthic community abundance and production, the abundance and production being a few to several times that of the macrobenthos (Giere 2009). Others, however, present different findings showing that carbon mineralisation pattern by the meiobenthos, as inferred from oxygen consumption, is far from clear-cut and seems to be habitat-dependent. On the one hand, de Assunção et al. (2010) demonstrated, in the southern North Sea, that—under oxic conditions—meiobenthic nematodes were as important as the macrobenthos in respiring available carbon, as measured by their contribution to the total sediment community oxygen consumption (SCOC). Other studies from coastal waters confirm the significance of the meiobenthos to the total community metabolism; for example, Opaliński et al. (2010) found the benthic meiofauna to be the second, after the zooplankton, major oxygen consumer in the shallow sandy littoral, responsible for 26 % of the total O_2 consumption there. On the other hand, when such measurements were performed for dysoxic conditions (de Assunção et al. 2010), the emerging pattern was different and meiofaunal contribution turned out to be far inferior to that of microorganisms. Relevant data from the deep sea are very scarce. The only direct measurements of deep-sea meiobenthos respiration rates were performed by Shirayama (1992) who, however, did not place his findings in the wider context of an entire benthic community, but concluded that the metabolic intensity of nematodes (universal meiobenthos dominants) was higher than that of other benthic taxa. Inferences as to the partitioning of oxygen consumption in the deep sea, based on total SCOC measurements, point toward microbial communities and meio- and macro-sized protists (Foraminifera), rather than metazoan meiofauna, as contributing the most to benthic respiration of the available carbon, particularly following events of phytodetritus input to the seafloor (e.g. Pfannkuche 1993; Woulds et al. 2007).

Despite the lack of complete clarity as to the actual contribution of the meiofauna to the benthic metabolism as outlined above, the meiobenthos has, over the years, become established as a useful (and interesting!) ecological category for benthic studies, particularly in view of its realised and potential value in providing insights into a wide range of biological and ecological questions. Such questions include, for instance, speciation and related phenomena (e.g. introgressive hybridization; Hummon 1977), physiological adaptations to stressful environmental conditions, particularly oxygen deficiency or anoxia (Danovaro et al. 2010), mechanisms of detoxication (Powell et al. 1979), symbiosis with bacteria (Van Gaever et al. 2006), responses to and utilization of organic enrichment (e.g. Austen and Widdicombe 2006), to mention just a few areas of biology and ecology that benefitted from particularly important insights provided by research based on meiobenthic organisms. A detailed discussion on the utility of meiobenthic research in various areas of marine science has been recently provided by Balsamo et al. (2012) and Urban-Malinga (2013).

Marine ecological research often addresses entire meiobenthic communities considered as assemblages of interacting individual components represented by the "major taxa" (see above), but responding to external drivers as a more or less coherent entity. The structure of this entity is described in terms of its total abundance as well as the abundance (density) and proportions of various component taxa (examined at various levels of taxonomic resolution) that make up the assemblage. Much less common, particularly with respect to the deep-sea, are studies in which meiobenthic taxonomy is more finely resolved, and the genera or even species representing individual major taxa are listed. Such a task is rendered difficult by the need of specialised knowledge on individual taxa. Moreover, to be adequately identified to a low taxonomic level, some taxa require observations of live specimens, a condition which seldom (and almost never during expeditions to the deep sea) can be satisfied if and when a study is of an exploratory nature and/or targets the entire meiobenthic assemblage. Therefore, although there is awareness of an immense taxonomic richness (diversity) of the meiobenthos (Giere 2009), the knowledge of this diversity is still greatly limited.

Whenever detailed taxonomic studies on the meiobenthos have been carried out, they resulted in descriptions of numerous new species and genera, or even taxa of as high a taxonomic rank as phylum (e.g. the Loricifera; Kristensen 1983). Such studies obviously contribute to enhancing the knowledge of marine biodiversity which is far from sufficient. The insufficiency of that knowledge has been particularly well demonstrated in publications dealing with deep-sea meiobenthos taxonomy (e.g. Boucher and Lambshead 1995) and in the debate on shallow-water *versus* deep-sea biodiversity (Gage 1996; Gray 1994; Gray et al. 1997).

On the other hand, the paucity of taxonomic information on small benthic animals (mainly the meiobenthos), coupled with the continuous demand for various environmental impact assessments based on or involving observations of benthic community responses to disturbance (e.g. Goodsell et al. 2009), has triggered a discussion among marine biologists and ecologists as to the taxonomic identification level sufficient for detection of changes in the marine environment and communities (Bett and Narayanaswamy 2014; Ferraro and Cole 1990; Herman and Heip 1988; Maurer 2000; Terlizzi et al. 2003; Warwick 1988). The discussion remains largely unresolved: for example, Bertasi et al. (2009) found the species- and genus-level identification of polychaetes to be important in discerning impacts, while impacts could be detected with bivalves identified to the family level alone. The discussion is of both theoretical and practical importance. To advance knowledge on the deep-sea diversity patterns and to resolve certain pertinent questions concerning, e.g. endemism *versus* cosmopolitanism of deep-sea organisms, their evolution, connectivity between areas, gene flow etc., the taxonomic knowledge should be brought down to the finest resolution possible, including molecular/genetic. The discussion is important also because the detailed taxonomic knowledge is required to address questions of ecosystem functioning and its relationship to biodiversity (Loreau 2000). For example, the question of species redundancy, that is whether a species can be eliminated from a community without much harm to the functioning of the latter (Fonseca and Ganade 2001), cannot be appropriately addressed without knowledge on the identity of the species present in an assemblage. So far, considering that such knowledge with regard to the shallow-water meiobenthos is still limited, discussion as to species' complementarity in the deep-sea meiofauna is not likely to be initiated soon. On the other hand, in the environmental survey practice where time and cost-effectiveness are at premium, determination of the taxonomic level which could be managed by generalist rather than specialist research personnel, and which could still provide meaningful answers on impacts (in a correctly designed study; cf. Green 1979) would be highly desirable.

In general, the meiobenthic communities, although very diverse, usually show a well-developed dominance structure, in which one or two major taxa (usually free-living nematodes and, frequently, harpacticoid copepods) dominate the assemblage. This is true also with respect to the deep sea, as will be shown in subsequent chapters.

For the knowledge on functioning of any assemblage of organisms to be complete, one has to have—in addition to information on the assemblage's taxonomic make-up—data on activities of different components of the assemblage and on interrelationships between individual populations. One aspect of the problem involves the natural temporal variability in an assemblage of organisms and populations. While variation in meiobenthic communities over time has been followed in numerous studies, at different temporal resolutions, in shallow-water environments (e.g. Coull 1985), there is very little information on changes in deep-sea meiobenthic assemblages at time scales lower than a few years. A vital contribution to elucidating temporal changes in the deep-sea meiobenthos has been provided by Kalogeropoulou et al. (2010). In their 11-year study carried out in the Porcupine Abyssal Plain (NE Atlantic, 4,850 m depth) involving 10 sampling events, they were able to reveal temporal changes on the scale of years and seasons in response to qualitative and quantitative changes in fluxes of organic matter to the seafloor. The study by Kalogeropoulou et al. (2010), conducted in a deep-sea area showing a strong seasonality in the organic matter supply to the seafloor and a massive increase in the density of a holothurian species, stresses the need for information on temporal variability in deep-sea meiobenthic communities. The absence of such information makes it difficult to disentangle a signal of natural variability from that induced by, e.g. an anthropogenic disturbance event.

Another aspect involves functional attributes of meiobenthic organisms and communities. Functional characteristics of meiobenthic communities, such as, e.g. trophic relationships within the meiofauna itself and between the meiofauna and other benthic compartments, community metabolism, matter and energy flow, etc., and consequently the role the meiobenthos plays in seafloor biocoenoses have been analysed using various approaches. They all require that information on functional characteristics of meiobenthic components be assembled, from observations and experiments, and analysed, frequently with the aid of multivariate mathematical techniques. This is the basis of the so-called Biological Trait Analysis (BTA) (Bremner et al. 2006; Schratzberger et al. 2007) which, developed for and applied mainly to the macrobenthos (Bremner et al. 2006) has been gaining popularity and seems to be a promising approach in studies of meiobenthic assemblages or their subsets such as, e.g. nematodes (Schratzberger et al. 2007; Alves et al. 2014). BTA involves assigning various taxonomic units (species, genera) to different functional groups, e.g. trophic types (guilds). Trophic types are distinguished based mainly on morphological characteristics of mouth parts of the animals concerned, from which their potential food or prey is inferred. The most common example with regard to the meiobenthos is the structure of the buccal cavity of free-living marine nematodes (Nematoda) (Jensen 1987; Wieser 1953; cf. Fig. 3.1 in Chap. 3). Recently, Alves et al. (2014) applied functional characteristics (BTA) of the meiobenthos (specifically, the nematodes) to detect an environmental change (deteriorating O_2 conditions). Trophic guild identification is aided by the analysis of gut contents (Jensen 1987) and feeding modes and food preferences (Berghe and Bergmans 1996; Hellwig-Armonies et al. 1991; Moens and Vincx 1997; Riera et al. 1996). Other morphological traits (size and shape of an individual, tail shape in the case of nematodes) have been used as well (e.g. Alves et al. 2014; Thistle et al. 1995). Other approaches helpful in elucidating functional aspects of meiobenthic community structure involve determination, in the field and/or in the laboratory, of biomass, metabolic rate (respiration rate; see above) (e.g. Herman and Heip 1982; Moens et al. 1996; Pamatmat and Findlay 1983), and production of individual taxa (e.g. Ceccherelli and Mistri 1991; Drgas 1999; Feller 1982; Fleeger and Palmer 1982; Herman and Heip 1985).

One of the important results of meiobenthic research to date has been the discovery that the meiofauna—both as an ecological entity and its individual components—is a sensitive indicator of environmental changes (Raffaelli 1982). Populations of various taxa (most often the meiobenthos dominants, i.e. nematodes and harpacticoids) have been used in laboratory experiments and field manipulations to determine impacts of various factors, both natural and anthropogenic, that disturb the natural equilibrium of the sedimentary environment in different water bodies (Austen et al. 1994; Essink and Romeyn 1994; Kovatch et al. 2000; Shaw et al. 1983). An obvious extension of studies of this kind is to include the meiobenthos into the suite of parameters regarded as indicators or response variables in biomonitoring of degradation-prone, or already perturbed, systems (Balsamo et al. 2012; Heip 1980; Keller 1986; Van Damme et al. 1984; Vincx and Heip 1987). Indeed, structural parameters of meiobenthic communities (overall abundance, proportions between various taxa) proved useful in elucidating effects of, e.g. eutrophication (Rokicka-Praxmajer et al. 1998), trace metal contamination (Austen and McEvoy 1997), oil pollution (Danovaro et al. 1995), or physical disturbance by both natural forces (Ólafsson and Elmgren 1991) and anthropogenic activities (Vanaverbeke et al. 2003), and also in monitoring the post-disturbance developments (Boucher 1985, Radziejewska, personal observations).

The utility of the meiobenthos in the assessment of environmental impacts has been amply documented in publications (such as those cited above) reporting on research conducted primarily in shallow waters, in relatively easily accessible coastal or shelf areas. The body of information accumulated from those areas is in sharp contrast with the paucity of corresponding data on meiobenthos of the largest depths of the oceans, the abyssal and the hadal. In view of proposals to use the deep seafloor for storing wastes, including nuclear (Angel and Rice 1996; Jahnke 1998; van der Loeff and Lavaleye 1986; Young and Richardson 1998), for CO_2 sequestration (Carman et al. 2004; Fleeger et al. 2006; Thistle et al. 2007a, b), and for extraction of the deep-sea mineral resources (see the Prologue), the need to augment the existing information on the deep-sea communities in general, and on the meiobenthos in particular, becomes all the more urgent. Also because structural parameters of meiobenthic communities are, as already mentioned, useful in marine monitoring, analysis of changes in the deep-sea meiobenthos in response to natural and man-mediated environmental alterations can bring interesting and important information on consequences of disturbance (van der Loeff and Lavaleye 1986). However, inferences that can be drawn from such analyses are greatly limited by the already mentioned insufficiency of data on the natural variability of deep-sea communities and habitats.

The awareness of the wide discrepancy between knowledge on shallow water communities and on those inhabiting deep oceanic bottoms, including the meiobenthos, has in recent decades led to intensified efforts aimed at bridging the gap. Those efforts have been seen in large-scale national and international research programmes such as, e.g. the Joint Global Ocean Flux Study (JGOFS; Fasham et al. 2001; Khripounoff et al. 1998), Census of Marine Life (CoML) and its CedaMar and COMARGE sections (Snelgrove 2010; McIntyre 2010) to name just a few. As a result, more and more data are being accumulated which hopefully allow to cross the "last frontier" of knowledge, as the deep-sea ecosystems have been termed (Lemonick 1995).

In view of the considerations presented above, this work was conceived as an attempt to provide a synopsis of the current knowledge on the meiobenthos of those abyssal areas of the Pacific which are likely to be targeted and impacted by future anthropogenic intervention, particularly by the development (mining) of mineral resources in the form of polymetallic (ferromanganese) nodules. This synopsis is intended as an aid in addressing the question if, and to what extent at the present stage of knowledge, the deep-sea meiobenthos can be used as a proxy in the assessment of potential consequences of anthropogenic disturbance for the sedimentary environment of abyssal areas.

This objective will be approached by, first, describing the environmental setting of the sub-equatorial north-east Pacific abyssal plain, with a particular attention paid to the Clarion-Clipperton Fracture Zone (CCFZ), an area constituting an immense polymetallic nodule field. This will be done in Chap. 2 which discusses the area's oceanographic features, environmental gradients, seafloor topography, sedimentary cover, and other components in the set of conditions defining the life space of the benthic communities, including the variability of those conditions and its scales. Attention will be also drawn to the objects of the anthropogenic activities planned in the area, i.e., nodule deposits. Subsequently, Chap. 3 will present an overview of the current knowledge on the meiobenthos of the CCFZ nodule field, including a discussion of methodological problems of the research, composition of the meiobenthos, diversity issues, and quantitative characteristics of distribution and spatio-temporal variability of the meiobenthic communities in the area. Chapter 4 will start with general considerations regarding the use of meiobenthos as a provider of response variables to seafloor disturbance, and will proceed by addressing the question of deep-sea mineral resources development *versus* the integrity of benthic communities. This part will provide an overview of field studies ("experiments") aimed at assessing the magnitude of potential impacts associated with polymetallic nodule development in the Pacific and the outcomes of using the meiobenthos-related variables as impact proxies. Finally, the Epilogue will conclude with a short summary of pertinent issues and perspectives for future work.

References

Alves A, Verissimo H, Costa MJ et al (2014) Taxonomic resolution and Biological Traits Analysis (BTA) approaches in estuarine free-living nematodes. Est Coast Shelf Sci 138:69–78

Angel MV, Rice TL (1996) The ecology of the deep ocean and its relevance to global waste management. J Appl Ecol 33:915–926

de Assunção FM, Vanaverbeke J, van Oevelen D et al (2010) Respiration partitioning in contrasting subtidal sediments: seasonality and response to a spring phytoplankton deposition. Mar Ecol 31:276–290

Austen MC, McEvoy AJ (1997) The use of offshore meiobenthic communities in laboratory microcosm experiments: response to heavy metal contamination. J Exp Mar Biol Ecol 211:247–261

Austen MC, Widdicombe S (2006) Comparison of the response of meio-and macrobenthos to disturbance and organic enrichment. J Exp Mar Biol Ecol 330:96–104

Austen MC, McEvoy AJ, Warwick RM (1994) The specificity of meiobenthic community responses to different pollutants: results from microcosm experiments. Mar Poll Bull 28:557–563

Balsamo M, Semprucci F, Frontalini F et al (2012) Meiofauna as a tool for marine ecosystem biomonitoring. In: Cruzado A (ed) Marine ecosystems. InTech, Rijeka. doi:10.5772/34423

Berghe WV, Bergmans M (1996) Differential food preferences in three co-occurring species of *Tisbe* (Copepoda, Harpacticoida). Mar Ecol Prog Ser 4:213–219

Bertasi F, Colangelo MA, Colosio F et al (2009) Comparing efficacy of different taxonomic resolutions and surrogates in detecting changes in soft bottom assemblages due to coastal defence structures. Mar Poll Bull 58:686–694

Bett BJ (2013) Characteristic benthic size spectra: potential sampling artefacts. Mar Ecol Prog Ser 487:1–6

Bett BJ, Narayanaswamy BE (2014) Genera as proxies for species α- and β-diversity: tested across a deep-water Atlantic-Arctic boundary. Mar Ecol. doi:10.1111/maec.12100

Boucher G (1985) Long term monitoring of meiofauna densities after the "Amoco Cadiz" oil spill. Mar Poll Bull 16:328–333

Boucher G, Lambshead PJD (1995) Ecological biodiversity of marine nematodes in samples from temperate, tropical and deep-sea regions. Conserv Biol 9:1594–1604

Bremner J, Rogers SI, Frid CLJ (2006) Matching biological traits to environmental conditions in marine benthic ecosystems. J Mar Sys 60:302–316

Carman KR, Thistle D, Fleeger JW et al (2004) Influence of introduced CO_2 on deep-sea metazoan meiofauna. J Oceanog 60:767–772

Ceccherelli VU, Mistri M (1991) Production of the meiobenthic harpacticoid copepod *Canuella perplexa*. Mar Ecol Prog Ser 68:225–234

Coull BC (1985) Long-term variability of estuarine meiobenthos: an 11 year study. Mar Ecol Prog Ser 24:205–218

Danovaro R, Fabiano M, Vincx M (1995) Meiofauna response to the "Agip Abruzzo" oil spill in subtidal sediments of the Ligurian Sea. Mar Poll Bull 30:133–145

Danovaro R, Dell'Anno A, Pusceddu A et al (2010) The first metazoa living in permanently anoxic conditions. BMC Biol 8(1):30

Drgas A (1999) Rola meiofauny w biocenozach dennych Zatoki Gdańskiej ze szczególnym uwzględnieniem wolnożyjących nicieni (Nematoda). PhD thesis, Sea Fisheries Institute, Gdynia

Drgas A, Radziejewska T, Warzocha J (1998) Biomass size spectra of near-shore shallow-water benthic communities in the Gulf of Gdańsk (Southern Baltic Sea). PSZN Mar Ecol 19:209–228

Duplisea DE, Drgas A (1999) Sensitivity of a benthic, metazoan, biomass size spectrum to differences in sediment granulometry. Mar Ecol Prog Ser 177:73–81

Duplisea DE, Hargrave BT (1996) Response of meiobenthic size-structure, biomass and respiration to sediment organic enrichment. Hydrobiologia 339:161–170

Essink K, Romeyn K (1994) Estuarine nematodes as indicators of organic pollution: an example from the Ems estuary (the Netherlands). Neth J Aquat Ecol 28:213–219

Fasham MJR, Baltiño BM, Bowles MC et al (2001) A new vision of ocean giobeochemistry after a decade of the Joint Global Ocean Flux Study (JGOFS). Ambio Spec Rep 10:4–30

Feller RJ (1982) Empirical estimates of carbon production from a meiobenthic harpacticoid copepod. Can J Fish Aquat Sci 39:1435–1443

Ferraro SP, Cole FA (1990) Taxonomic level and sample size sufficient for assessing pollution impacts on the Southern California Bight macrobenthos. Mar Ecol Prog Ser 67:251–261

Fleeger JW, Palmer MA (1982) Secondary production of the estuarine, meiobenthic copepod *Microarthridion littorale*. Mar Ecol Prog Ser 7:157–162

Fleeger JW, Carman KR, Weisenhorn PB et al (2006) Simulated sequestration of anthropogenic carbon dioxide at a deep-sea site: effects on nematode abundance and biovolume. Deep Sea Res I 53:1135–1147

Fonseca CR, Ganade G (2001) Species functional redundancy, random extinctions and the stability of ecosystems. J Ecol 89:118–125

Gage JD (1996) Why are there so many species in deep-sea sediments? J Exp Mar Biol Ecol 200:257–286

Giere O (2009) Meiobenthology. The microscopic fauna in aquatic sediments, 2nd edn. Springer, Berlin

Goodsell PJ, Underwood AJ, Chapman MG (2009) Evidence necessary for taxa to be reliable indicators of environmental conditions or impacts. Mar Poll Bull 58:323–331

Gray JS (1994) Is deep-sea species diversity really so high? Species diversity of the Norwegian continental shelf. Mar Ecol Prog Ser 112:205–209

Gray JS, Poore GCB, Ugland KI et al (1997) Coastal and deep-sea benthic diversities compared. Mar Ecol Prog Ser 159:97–103

Green RH (1979) Sampling design and statistical methods for environmental biologists. Wiley & Sons, New York

Heip C (1980) Meiobenthos as a tool in the assessment of marine environmental quality. Rapp Proc-verb Réun 179:182–187

Hellwig-Armonies M, Armonies W, Lorenzen S (1991) The diet of *Enoplus brevis* (Nematoda) in a supralittoral salt marsh of the North Sea. Helgol Meeresunters 45:357–371

Herman PMJ, Heip C (1982) Growth and respiration of *Cyprideis torosa* Jones, 1850 (Crustacea, Ostracoda). Oecologia (Berl) 54:300–303

Herman PMJ, Heip C (1985) Secondary production of the harpacticoid copepod *Paronyhocamptus nanus* in a brackish-water habitat. Limnol Oceanog 30:1060–1066

Herman PMJ, Heip C (1988) On the use of meiofauna in ecological monitoring: who needs taxonomy? Mar Poll Bull 19:665–668

Hummon WD (1977) Introgressive hybridization between two intertidal species of *Tetranchyroderma* (Gastrotricha, Thaumastodermatidae) with the description of a new species. Mikrofauna Meeresbodens 61:113–136

Jahnke R (1998) Geochemical impacts of waste disposal on the abyssal seafloor. J Mar Sys 14:355–375

Jensen P (1987) Feeding ecology of free-living aquatic nematodes. Mar Ecol Prog Ser 35:187–196

Kalogeropoulou V, Bett BJ, Gooday AJ et al (2010) Temporal changes (1989–1999) in deep-sea metazoan meiofaunal assemblages on the Porcupine Abyssal Plain, NE Atlantic. Deep Sea Res II 57:1383–1395

Keller M (1986) Structure des peuplements méiobenthiques dans le secteur pollué par le rejet en mer de l'égout de Marseille. Ann Inst Oceanogr 62:2–36

Khripounoff A, Vangriesheim A, Crassous P (1998) Vertical and temporal variations of particle fluxes in the deep tropical Atlantic. Deep-Sea Res I 45:293–316

Kovatch CE, Schizas NV, Chandler GT et al (2000) Tolerance and genetic relatedness of three meiobenthic copepod populations exposed to sediment-associated contaminant mixtures: role of environmental history. Environ Toxicol Chem 19:912–919

Kristensen RM (1983) Loricifera, a new phylum with aschelminthes characters from the meiobenthos. Zeitschr Zool Syst Evolutionsforsch 21:163–180

Lemonick MD (1995) The last frontier. Time Int 146(7):36–44

Loreau M (2000) Biodiversity and ecosystem functioning: recent theoretical advances. Oikos 91:3–17

Mare MF (1942) A study of a marine benthic community with a special reference to the microorganisms. J Mar Biol Ass UK 25:517–554

Maurer D (2000) The dark side of taxonomic sufficiency (TS). Mar Poll Bull 40:98–101

McIntyre AD (1969) Ecology of marine meiobenthos. Biol Rev Cambridge Phil Soc 44:245–290

McIntyre A (ed) (2010) Life in the world's oceans: diversity, distribution, and abundance. Wiley-Blackwell, Chichester

Moens T, Vincx M (1997) Observations on the feeding ecology of estuarine nematodes. J Mar Biol Ass UK 77:211–227

Moens T, Vierstraete A, Vanhove S et al (1996) A handy method for measuring meiobenthic respiration. J Exp Mar Biol Ecol 197:177–190

Ólafsson E, Elmgren R (1991) Effects of biological disturbance by benthic amphipods *Monoporeia affinis* on meiobenthic community structure: a laboratory approach. Mar Ecol Prog Ser 74:99–107

Opaliński KW, Maciejewska K, Urban-Malinga B et al (2010) The oxygen fluxes of sandy littoral areas: quantifying primary and secondary producers in the Baltic Sea. Mar Poll Bull 61:211–214

Pamatmat MM, Findlay S (1983) Metabolism of microbes, nematodes, polychaetes, and their interactions in sediment, as determined by heat flow measurements. Mar Ecol Prog Ser 11:31–38

Pfannkuche O (1993) Benthic response to the sedimentation of particulate organic matter at the BIOTRANS station, 47°N, 20°W. Deep Sea Res II 40:135–149

Powell EN, Crenshaw MA, Rieger RM (1979) Adaptations to sulfide in the meiofauna of the sulfide system. I. ^{35}S-sulfide accumulation and the presence of a sulfide detoxification system. J Exp Mar Biol Ecol 37:57–76

Raffaelli D (1982) An assessment of the potential of major meiofauna groups for monitoring organic pollution. Mar Envir Res 7:151–164

Riera P, Gremare A, Blanchard G (1996) Food sources of intertidal nematodes in the Bay of Garennes-Oleron (France), as determined by dual stable isotope analysis. Mar Ecol Prog Ser 142:303–309

Rokicka-Praxmajer J, Radziejewska T, Dworczak H (1998) Meiobenthic communities of the Pomeranian Bay (southern Baltic): effects of proximity to river discharge. Oceanologia 40:243–260

Schratzberger M, Warr K, Rogers SI (2007) Functional diversity of nematode communities in the southwestern North Sea. Mar Envir Res 63:368–389

Schwinghamer P (1981) Characteristic size distributions of integral benthic communities. Can J Fish Aquat Sci 38:1255–1263

Shaw KM, Lambshead PJD, Platt HM (1983) Detection of pollution-induced disturbance in marine benthic assemblages with special reference to nematodes. Mar Ecol Prog Ser 11:195–202

Shirayama Y (1992) Respiration rates of bathyal meiobenthos collected using a deep-sea submersible SHINKAI 2000. Deep-Sea Res 39:781–788

Snelgrove PV (2010) Discoveries of the census of marine life. Cambridge University Press, Cambridge

Terlizzi A, Bevilacqua S, Fraschetti S et al (2003) Taxonomic sufficiency and the increasing insufficiency of taxonomic expertise. Mar Poll Bull 46:556–561

Thiel H (1983) Meiobenthos and nanobenthos of the deep sea. In: Rowe G (ed) Deep-sea biology. The Sea, 8th edn. Wiley-Interscience, New York

Thistle D, Lambshead PJD, Sherman KM (1995) Nematode tail-shape groups respond to environmental differences in the deep sea. Vie Milieu 45:107–115

Thistle D, Sedlacek L, Carman KR et al (2007a) Emergence in the deep sea: Evidence from harpacticoid copepods. Deep-Sea Res I 54:1008–1014

Thistle D, Sedlacek L, Carman KR et al (2007b) Exposure to carbon dioxide-rich seawater is stressful for some deep-sea species: an *in situ*, behavioral study. Mar Ecol Prog Ser 340:9–16

Urban-Malinga B (2013) Meiobenthos in marine coastal sediments. Geol Soc London, Spec Publ 388. doi:10.1144/SP388.9

Van Damme D, Heip C, Willems KA (1984) Influence of pollution on the harpacticoid of two North Sea estuaries. Hydrobiologia 112:143–160

van der Loeff MMR, Lavaleye MSS (1986) Sediments, fauna and the dispersal of radionuclides at the N.E. Atlantic dumpsite for low-level radioactive waste. Report of the Dutch DORA Program, Netherlands Institute for Sea Research, Texel

Van Gaever S, Moodley L, De Beer D et al (2006) Meiobenthos at the Arctic Håkon Mosby Mud Volcano, with a parental-caring nematode thriving in sulphide-rich sediments. Mar Ecol Prog Ser 321:143–155

Vanaverbeke J, Steyaert M, Vanreusel A et al (2003) Nematode biomass spectra as descriptors of functional changes due to human and natural impact. Mar Ecol Prog Ser 249:157–170

Vincx M, Heip C (1987) The use of meiobenthos in pollution monitoring studies. A review. ICES, C.M. E:33

Warwick RM (1984) Species size distributions in marine benthic communities. Oecologia (Berl) 61:32–41

Warwick RM (1988) The level of taxonomic discrimination required to detect pollution effects on marine benthic communities. Mar Poll Bull 19:259–268

Warwick RM, Collins NR, Gee JM et al (1986) Species size distributions of benthic and pelagic Metazoa: evidence for interaction? Mar Ecol Prog Ser 34:63–68

Wieser W (1953) Die Beziehung zwischen Mundhohlengestalt, Ernahrungsweise und Vorkommen bei freilebenden marinen Nematoden. Ark Zool 4:439–484

Woulds C, Cowie GL, Levin LA et al (2007) Oxygen as a control on sea floor biological communities and their roles in sedimentary carbon cycling. Limnol Oceanog 52:1698–1709

Young DK, Richardson MD (1998) Effects of waste disposal on benthic faunal succession on the abyssal seafloor. J Mar Sys 14:319–336

Chapter 2
Characteristics of the Sub-equatorial North-Eastern Pacific Ocean's Abyss, with a Particular Reference to the Clarion-Clipperton Fracture Zone

Abstract The deep seafloor, i.e. seabed areas at depths exceeding 800–1,300 m, cover about 88 % of the world ocean's bottom. The most extensive areas represent the 3,000–6,000 depth range and include abyssal plains (depths > 4,000 m) the largest of which is the abyssal plain of the Pacific Ocean. The water column overlying it consists of a number of layers differing in their major characteristics. The most characteristic layers include that encompassing the oxygen minimum zone (OMZ, 100–1,000 m depth range in the Pacific) and the near-bottom layer, directly impinging on the seafloor. Once considered extremely stable, the near-bottom layer is now known to be prone to hydrodynamic effects such as tides and currents. The latter are generally weak, but periods of intensified current activity are not infrequent. The water column effects influencing the abyssal seafloor include also the transmission of the wind-generated surface physical energy down to the bottom (the "benthic storms") on the one hand and sedimentation of surface-produced organic matter on the other. Both the benthic storms and organic matter deposition are known to be periodically, or episodically, intensified, thus contributing to natural environmental variability in the abyss. The Pacific abyssal plain sedimentary cover is mostly biogenic in origin. A characteristic part of the Pacific abyss is a huge (about 2 million km^2) polymetallic nodule field within the NE sub-equatorial seafloor area experiencing low sedimentation rates and constrained by the Clarion and Clipperton fractures (the Clarion-Clipperton Fracture Zone, CCFZ). CCFZ extends sub-latitudinally along about 4,200 km, its surface inclining slightly westwards with depths in the east-west direction from about 4,000 to about 5,400 m. The seafloor, although generally flat, does show (particularly in the eastern part) distinct topographic features which are volcanic in origin. The relatively thin (50–200 m) sedimentary cover is formed by recent biogenic sediments (siliceous ooze). The major characteristic of the area is the presence of polymetallic nodule deposits. The nodules, occurring at abundances frequently exceeding 10 kg/m^2, are mostly exposed on the sediment surface, but some are also embedded or buried in the sediment. The nodules, in addition to the occasionally occurring larger hard-rock fragments of cobalt-rich ferromanganese crusts, add to the deep-sea habitat heterogeneity and themselves constitute both a unique

© The Author(s) 2014
T. Radziejewska, *Meiobenthos in the Sub-equatorial Pacific Abyss*,
SpringerBriefs in Earth System Sciences, DOI 10.1007/978-3-642-41458-9_2

deep-sea habitat, of a great interest to marine ecologists, and an important mineral resource, of a great appeal to the marine mining community the size of which has been recently growing considerably.

Keywords Abyssal depths · Clarion-Clipperton Fracture Zone · Hydrography · Sedimentation · Currents · Bottom topography · Sedimentary cover · Polymetallic nodules

The deep sea is defined as the part of the oceanic water column and seafloor that extends below the permanent oceanic thermocline, i.e. below depth contours of 800–1,300 m (Gage and Tyler 1991; Ramirez-Llodra et al. 2010). That part of the oceanic floor covers about 88 % of the world ocean's floor surface area. The most extensive deep sea bottom is located between the depths of 3,000 and 6,000 m, and generally represents fairly flat surfaces, called the abyssal plains (Angel and Rice 1996, Mantyla and Reid 1983). The largest among them is the abyssal plain of the Pacific Ocean (Fig. 2.1), its part located in the sub-equatorial north-eastern region of the ocean being the major focus of the description below.

Environmental conditions in the abyss have been described in detail in a number of studies important for a biologist interested in environmental effects of anthropogenic interventions on deep-sea communities. The most pertinent of those publications include the papers by Menzies (1965), Gage and Tyler (1991), Tyler (1995), and Hannides and Smith (2003). As well as providing important overviews of the deep-sea environment, the publications mentioned, the earliest and the most recent of which span a period of almost 40 years, supply evidence on how the human perception of the deep-sea environment changed as research methods and techniques were improving and more and more large-scale study programmes were being implemented.

Fig. 2.1 The spatial extent of the Pacific abyssal plain; the *shaded area* shown indicates seafloor at depths exceeding 3,500 m; drawing made using the ODV software (Schlitzer 2004)

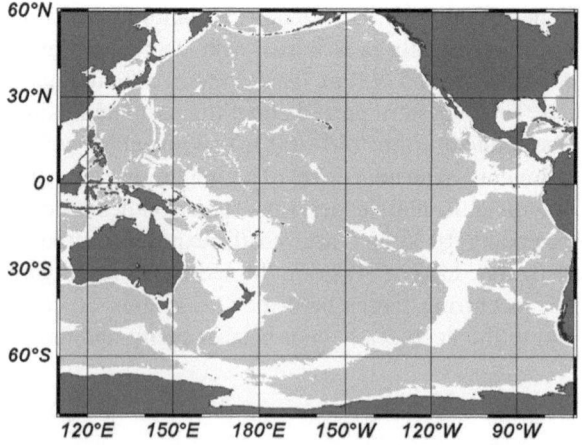

It is now recognised that the abyssal seafloor is influenced by phenomena and processes occurring in the surface waters which are themselves affected by atmospheric forcing agents—atmospheric circulation, wind stress, precipitation, evaporation, and energy fluxes (Amador et al. 2006). Phenomena such as eddies and mesoscale processes, El Niño Southern Oscillation (ENSO), and interdecadal variability and climate change (Willett et al. 2006; Wang and Fiedler 2006; Mestas-Nunez and Miller 2006) produce effects which, transmitted down the water column and modified along the way, influence the deep seafloor.

The water column overlying the abyssal seafloor shows a number of characteristic layers, differing in their physical, chemical, and biological variables. The most notable of those layers include the so-called oxygen minimum zone (OMZ) and the bottom (near-bottom) water, both of potential (OMZ) and actual (near-bottom water) importance for the benthic communities. The oxygen minimum zone in the eastern tropical Pacific is very pronounced; it occurs generally within 100–1,000 m (Hannides and Smith 2003; Wishner et al. 1995) and does not extend to the abyssal depths.

Another important water layer is the near-bottom one. The origins of the Pacific near-bottom water can be traced to the surface layer of the Antarctic region. As a result of complex interactions that involve mixing, sinking, and advection, that surface water sinks towards the bottom to form the Antarctic Bottom Water (AABW) in the Weddell Sea (Tsuchiya and Talley 1996). Its branches move northwards and penetrate into the Atlantic, Indian, and Pacific Oceans, the penetration routes being restricted and modified by topographic structures on the bottom, such as the East Pacific Rise. According to the pattern described in detail by Tsuchiya and Talley (1996), AABW moves along a major pathway towards the NE basin of the Pacific which is a route that carries the cold Antarctic water from the Equator to the north-east and eastwards via the Horizon Passage (~19°N; 169°W) at the northern tip of the Line Islands Ridge and via the Clarion Passage (~12°N; 165°W) in the western part of the Clarion Fracture. A portion of that stream flows eastwards through the northern part of the Clipperton Fracture. Bottom water (at depths exceeding 4,000 m) in the Pacific is formed by the Lower Circumpolar Water (LCPW), a mixture of AABW and the North Atlantic Deep Water (NADW) formed in the northern North Atlantic (Fiedler and Talley 2006). As described by Fiedler and Talley (2006), LCPW flows into the eastern tropical Pacific from the Northeast Pacific Basin along the western flank of the East Pacific Rise (105–110°W), at the end of a circuitous path in a northward deep western boundary current through the Central Pacific Basin and then as an eastward branch between the Hawaiian and Line Islands (Johnson and Toole 1993).

In many areas of the Pacific, including its eastern flanks, the hydrodynamic energy of currents is enhanced by hydrothermal venting (Tivey et al. 2002). Both the currents and the venting are recognised sources of a considerable variability in the oceanographic regime in the deep sea. Irrespective of that variability, however, the major trait of deep-sea bottom water masses is a fairly narrow range of changes of their major hydrographic characteristics in individual regions: over vast areas of the seafloor, the temperature varies from about 4 °C to about −1 °C, while the mean salinity is 34.8 ± 0.3 psu (dropping to the minimum mean value of 34.69 psu in the NE and eastern part of the Pacific). The bottom water masses overlying the sub-equatorial NE Pacific seafloor are old (in terms of the amount of time since their contact with the surface water layers), but their dissolved oxygen content is close to saturation. Although, as AABW moves away from the centre of its formation, the dissolved oxygen pool becomes gradually depleted due to metabolic processes (Mantyla and Reid 1983; Tsuchiya and Talley 1996), it does not drop below the level of 2 ml l^{-1}, a threshold critical for the survival of benthic communities (Rabalais et al. 2010).

Irrespective of the narrow ranges of oceanographic variables mentioned above, certain attributes of the deep sea show oceanographic gradients and manifestations of physical variability that strongly impinge upon biological communities of both the water column and the deep seafloor. Particularly notable in this respect are changes in hydrostatic pressure in the water column. Depth-dependent changes in hydrostatic pressure represent the longest continuous environmental gradient on Earth, the hydrostatic pressure increasing by 10^5 Pa with each 10 m depth increase (Gage and Tyler 1991). The importance of the gradient for physiology and ecology of deep-sea organisms is evident from a number of studies. The most important earlier publications include a collection of papers edited by Hochachka (1976). To cite a more recent work, Young and Tyler (1993) referred to hydrostatic pressure as a factor controlling the vertical distribution of various echinoid species. The hydrostatic pressure gradient has also been invoked to explain depth-related changes in diversity and density of benthic organisms: for example, Tyler (1995) contended that the hydrostatic pressure of about 6×10^7 Pa (equal to that prevalent at the depth of about 6,000 m) is a clear bathymetric boundary for the fauna. A number of higher taxa such as decapods, sea anemones, and echinoids are absent at depths exceeding 6,000 m, while other taxa, particularly holothurians and polychaetes, occur there at relatively high densities. Barophilic adaptations are known also among microorganisms (Fang et al. 2000; Tholosan et al. 1999).

In addition, the hydrostatic pressure affects calcium carbonate solubility and thus can indirectly influence the deep-sea fauna. The carbonate compensation depth (CCD), i.e. the depth at which calcium carbonate is no longer solid and becomes dissolved, lies at different depths in various oceans and averages globally about 4,500 m (Coxall et al. 2005). In the eastern Pacific, it is shallower than in other oceans, being found at about 3,500 m (Nienstedt and Arnold 1988; Seibold and Berger 1993; Tyler 1995) to drop to depths larger than 5,000 m in the western part of the CCFZ (Kotlinski, pers. comm.). The hydrostatic pressure-mediated calcium carbonate solubility is important for carbonate enrichment of the deep seafloor and participates in controlling the chemical composition of sediments at those depths (Brown et al. 1989). It is also important for the composition of abyssal benthic communities by greatly restricting the distribution of calcareous shell-bearing organisms, e.g. calcareous foraminiferans (Gooday 1994).

The abyssal seabed topography results from interaction between continental plate spreading characteristics and sedimentation of organic and inorganic particles from shallower depths (Kotliński 1999; Whitmarsh et al. 1996). Although abyssal plains are relatively flat compared to the continental margins and slopes flanking them, modern techniques of abyssal topography exploration (e.g. multibeam surveys) show that the abyssal relief can be rendered complex by the presence of guyots, seamounts, and underwater volcanoes that often occur in chains and as seamount ridges, as well as by a multitude of smaller-scale topographic features (Kotliński, pers. comm.). Moreover, since the abyssal plains are intersected by mid-oceanic ridges, their surfaces are fractured in numerous places. The areas away from fractures and ridges show very gentle slope angles (on the order of 1:1,000) (Gage and Tyler 1991).

The abyssal plain sedimentary cover is mainly biogenic in origin: it forms as a result of sedimentation of remnants of pelagic fauna and flora which, once on the seafloor, are subjected to diagenetic processes of various intensity, depending on temperature and pressure (Brown et al. 1989; Menzies 1965, Seibold and Berger 1993; Tyler 1995; Whitmarsh et al. 1996). Rates of sedimentation and sediment accumulation vary between and within the oceans, depending largely on productivity of the surface waters: for example, the sediment accumulation rate in the northern Atlantic is estimated at 3.0–5.0 mm/kyear against the estimated 1.6 mm/kyear in NE Pacific (Kotliński and Rühle 1998). The deposited particles give rise to the so-called pelagic sediments, very finely grained, forming a highly hydrated surface layer which overlays the hard basalts of the oceanic fundament (Brown et al. 1989; Kotliński 1999; Seibold and Berger 1993). The thickness of those sediments depends on the crust age and sedimentation rate; the sedimentary cover in areas away from spreading centres can be several hundred metres thick. Such sediments cover vast expanses of abyssal plains, imparting on them a characteristically monotonous landscape shown in underwater images of the abyssal zone (Fig. 2.2). The soft sediment is thus the basic substratum type in the abyssal. Hard substrata, particularly in the form of basaltic rock outcrops, are much less common, but exist nonetheless. Hard substrata are mainly restricted to regions of

Fig. 2.2 A view of the Pacific abyssal plain in the eastern part of the Clarion-Clipperton Fracture Zone (CCFZ) (*Photo* courtesy of IOM)

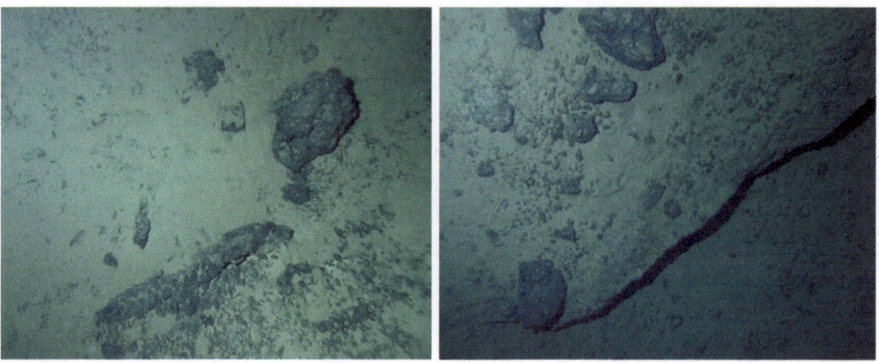

Fig. 2.3   A nodule-covered seafloor with crust fragments in the CCFZ (*Photo* courtesy of IOM)

recent volcanic activity and areas of weak and very weak sedimentation (Hannides and Smith 2003). In the latter, the hard substratum is provided primarily by poly-metallic (ferromanganese) nodule deposits and associated polymetallic (cobalt-rich ferromanganese) crust fragments, present in certain areas (Brown et al. 1989; Hein and Petersen 2013; Kotliński 1999; Seibold and Berger 1993) (Fig. 2.3).

The seaflooor topography and the surficial sediment layer are involved in com-plex interactions between major factors governing the conditions of life of benthic communities (Snelgrove and Butman 1994), and are affected by hydrodynamic energy of the near-bottom water layer. Vectors of that energy are the near-bottom currents, one of the least known physical forces in the deep sea (Gage and Tyler 1991). Information on the near-bottom current pattern in the abyssopelagic regions is very scant due to the absence, in most part of the world's oceans, of data from direct current meter measurements. It is assumed that the currents active in that region are controlled by forces resulting from thermohaline and tidal phenomena. The thermohaline circulation is a result of the formation of dense water masses around the Antarctic and their sinking to deeper layers, coupled with movement towards the Equator and north of it, along corridors formed by bottom topography (Tsuchiya and Talley 1996). The water flow resulting from thermohaline variabil-ity is modified by a tidal component; its strongest manifestation is a semi-diurnal shift in current direction. There is evidence showing that the pattern of those inter-nal deep-sea tides may also reflect spring-neap oscillations. In addition, the result-ant near-bottom current has been observed to change its direction periodically, or even to reverse it, as a result of kinetic energy being transferred from the surface to the bottom of the water column in eddies forming on the surface (Willett et al. 2006) and spanning the entire water column down to the near-bottom layers (Kontar and Sokov 1994; Sokov and Demidova 1992). It is also assumed that this transfer of kinetic energy contributes significantly to the motion of the deep sea water masses. Mesoscale (50–200 km) eddies carry an amount of energy up to 100 times that of the background level. Such eddies are formed as a result of wind action upon the ocean's surface; their energy is transferred to the abyssopelagial where it may periodically change current direction and increase current velocity. The process, termed a "benthic storm" (Aller 1997; Brown et al. 1989; Kontar and Sokov 1994), can destabilise the sediment surface and erode it, thereby directly influencing the structure of benthic communities (Aller 1997; Thistle 1998; Thistle and Levin 1998; Thistle et al. 1999).

The basic source of energy for abyssal benthic communities is the biological production in the water column, particularly in its surface layers. A significant portion of that production is the organic matter formed by primary producers, the phytoplankton. Under certain conditions, the phytal material produced in the top water layers may sediment all the way down to the abyssal depths and reach the seafloor there as the relatively undegraded phytodetritus (e.g. Billett et al. 1983; Pfannkuche et al. 1999; Scharek et al. 1999; Thiel et al. 1988/1989). Usually, however, while sinking, the organic material undergoes various transformations: it may be subject to degradation, decomposition, and partial dissolution. It is invariably colonised by microorganisms and may be enveloped by extracellular polymers they produce, which enhances scavenging of other particulate material in the water column and the formation of macroaggregates known as the "marine snow" (Alldredge and Silver 1988). Thus, some of the transformations may increase the particle sinking rate (Angel 1984; Brown et al. 1989; Fowler and Knauer 1986). Data provided by analyses of sediment trap contents show that as much as 1–7 % of the surface production may reach the abyssal bottom (Pfannkuche et al. 1999; Tyler 1995). Another form of organic matter flux to deep-sea sediments occurs as rather ephemeral, but important for local organic enrichment, events of sinking of large organic remains (dead macrophytes, fish, cephalopods or whales) onto the deep seafloor [cf. research on whale falls reviewed by Smith and Baco (2003) and Smith et al. (1998)].

Geochemical conditions in the sediment were reported to covary with the surface primary production rates and the particulate organic carbon (POC) fluxes (Hannides and Smith 2003). The geochemical conditions are reflected by profiles of dissolved oxygen in the sediment, oxygen penetration generally increasing with increasing distance from the Equator (Hammond et al. 1996) as a result of decreasing oxygen metabolic consumption within the sediment.

The picture of abiotic components of abyssal ecosystems was initially painted by studies leading to the milestone paper by Menzies (1965) who described a relative homogeneity of salinity, temperature, and dissolved oxygen content of the water masses overlying vast expanses of the abyssal seafloor. This gave rise to a widely held notion, a paradigm of sorts, that abyssal ecosystems are extremely stable, the only manifestations of instability involving the hydrostatic pressure and the occasional near-bottom turbidity currents. That deeply rooted conviction was challenged within the past two decades or so, as data collected during large-scale oceanographic research programmes such as the World Ocean Circulation Experiment (WOCE) (Tsuchiya and Talley 1996) or the Joint Global Ocean Flux Study (JGOFS) (Stoecker et al. 1996) began to accumulate and reveal a different picture. More and more frequently, evidence of variability—along different spatial and temporal scales—in a number of factors shaping the abyssal environment has been provided. The variability is manifested as, e.g. diel changes in near-bottom current patterns brought about by deep-sea tides, or by important disturbances in the hydrodynamics due to the downward transport of the surface wind gyre energy (Kontar and Sokov 1994, Sokov and Demidova 1992). Biotic parameters, too, are subject to variability which is largely seasonal, as demonstrated by studies on vertical POC fluxes from the near-surface water layers. Based on POC fluxes reported in the literature, Hannides and Smith (2003) divided the NE Pacific abyssal plain into three zones: the eutrophic abyss (from the Equator up to 5°N), with the POC flux of about 1–2 g C/m^2 year; the mesotrophic abyss (from 5°N to 15°N), with the POC flux of about 0.5–1.6 g C/m^2 year; and the oligotrophic abyss (underlying the central North Pacific gyre), with POC fluxes typically lower than 0.5 g C/m^2 year. Variation in the quality and rate of the POC flux was reported to affect the sediment bio- and geochemistry (Hammond et al. 1996) as well as the functioning and structure of benthic communities (Kalogeropoulou et al. 2010; Tyler 1988).

In this context, sedimentation of the phytodetritus, mentioned above, seems to be extremely important. Effects of that sedimentation on geo- and bio-chemical processes in the abyssal sediment as well as its significance in controlling variation in food resources, the structure and functioning of benthic communities, and physiological processes of abyssal organisms have been vigorously pursued over the recent several years (Fileman et al. 1998; Kalogeropoulou et al. 2010; Khripounoff et al. 1998; Pfannkuche et al. 1999; Smith et al. 1996, 1997). Biological effects of phytodetritus input to the abyssal seafloor will be discussed in Chap. 3.

The abyssal biology may also reflect variability effected on longer temporal and wider spatial scales, e.g. changes in surface biological production brought about by global climatic phenomena such as the El Niño Southern Oscillation (ENSO) (Gage and Tyler 1985; Neira et al. 2001; Pennington et al. 2006; Tyler 1995). For example, Pennington et al. (2006) cite studies which reported the warm phases of ENSO, the El Niños, to have several-fold reduced nutrient supply, with a concomitant decrease in primary production and water chlorophyll *a* content in the eastern tropical Pacific, while the cool phase of ENSO called La Niña produced an opposite effect in the form of a massive bloom with predictably dramatic biological consequences.

As already said, of a particular importance for this book is the Clarion-Clipperton Fracture Zone (CCFZ), a part of the sub-equatorial NE Pacific's abyssal plain (Fig. 2.4), and particularly its eastern part housing the claim area of Interoceanmetal Joint Organization (IOM) (cf. Fig. 3; Kotlinski 1998b). As CCFZ is a site of extensive and abundant polymetallic nodule deposits, its geological set-up and the characteristics important from the standpoint of future development of the deposits have already been, and are, vigorously explored and amply documented (e.g. Andreev and Gramberg 1998; Barash et al. 2000; IOM 1993; Kotliński 1998a, b, 1999; Tkatchenko and Stoyanova 1998; Tkatchenko et al. 1997). CCFZ covers about 2 million km^2 of the abyssal plain seafloor in the sub-equatorial north-eastern part of the Pacific, flanked to the north by the Clarion Fracture and by the Clipperton Fracture to the south (Fig. 2.4). The area extends sub-latitudinally along about 4,200 km and is about 300–900 km wide. The area's

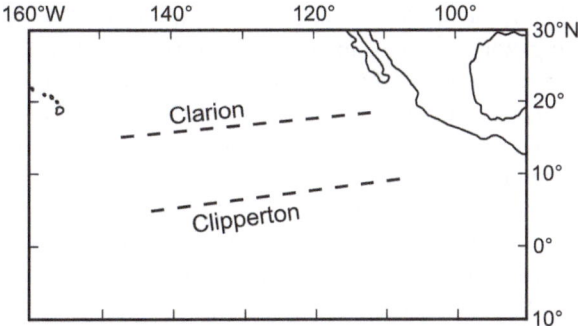

Fig. 2.4 A schematic of locations of the Clarion and Clipperton Fractures in the sub-equatorial NE Pacific

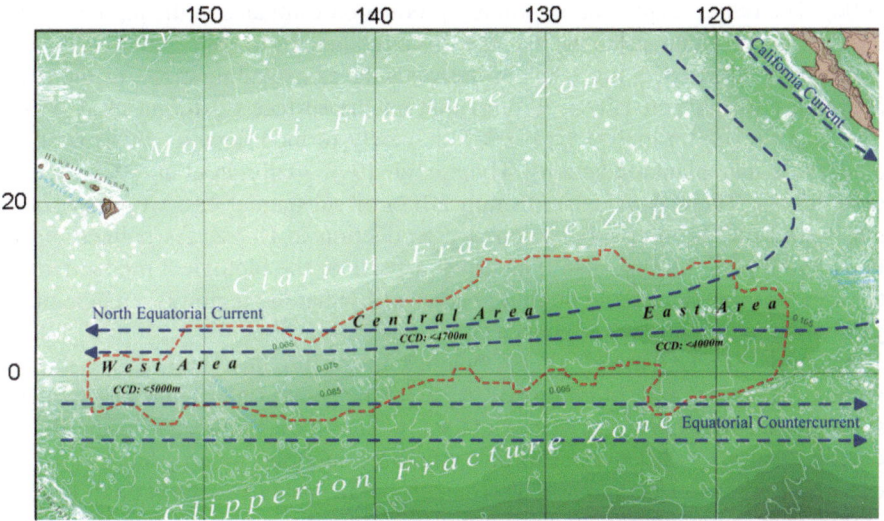

Fig. 2.5 Oceanographic features of the sub-equatorial NE Pacific; CCFZ contoured by *dashed line* in *red* (Drawing courtesy of R. Kotliński)

hydrography is shaped by a system consisting of the westward-flowing Northern Equatorial Current and by the eastern-flowing Northern Equatorial Countercurrent (Fig. 2.5). The currents form a divergence system, very complicated due to its seasonal variability and local and regional wind eddies (Fiedler and Talley 2006; Sokov and Demidova 1992; Willett et al. 2006). In spring and summer, the northern boundary of the Equatorial Countercurrent reaches to 12–13°N, to shift southwards to 10°N in autumn and winter (Ozturgut et al. 1978; Tkatchenko and Stoyanova 1998).

The CCFZ water column structure reveals the presence of 5 water layers differing in their temperature and salinity (IOM 1993; Johnson and Toole 1993;

Tkatchenko and Stoyanova 1998; Tkatchenko et al. 1997; Wijffels et al. 1996). The upper mixing zone extends to about 50–150 m (Demidova 1999), while the oxygen minimum zone (OMZ) encompasses the depth range of 300–500 m (Ozturgut et al. 1978); in the eastern part of the CCFZ, the OMZ was recorded to span the depth range of 120–2,500 m (Kotliński et al. 1996), with dissolved oxygen contents dropping to well below 1 ml O_2 /l. The near-bottom layer shows temperature and salinity ranges of 1.3–1.6 °C and 34.4–34.88 psu, respectively, the dissolved oxygen content averaging 3.6–3.8 ml O_2 /l (Kotliński et al. 1996).

The recent literature reviewing satellite observations-based (SeaWiFS; http://seawifs.gsfc.nasa.gov) data on chlorophyll a distribution (Morgan 2000; Pennington et al. 2006) shows the chlorophyll a values in the CCFZ to be generally rather low (on the order of 0.1–0.25 mg/m^3), but higher than in the areas north of CCFZ (cf. Fig. 2.5). In addition, SeaWiFS chlorophyll estimates seem to indicate a seasonal cycle, with higher values in the boreal winter (Pennington et al. 2006). The rather scant data on primary production estimates in the top part of the CCFZ water column show the average daily production to vary widely from 20.4 (Kolosov 1988) to 133 mg C/m^2 in the water layer extending down to the depth at which the light intensity is 1 % of that on the surface (Ozturgut et al. 1978). The major (even up to 100 %) part of the newly formed organic matter is attributed to the nanoplankton activity (Hyun et al. 1998; Ozturgut et al. 1978; Semina et al. 1997). There is evidence (Smith 2013) of substantial east-west and south-north gradients in the ovelying primary production in CCFZ. As pointed out by Pennington et al. (2006), many important aspects of the biological productivity of the area remain unclear, although it is noteworthy that the region is a major site of yellowfin tuna occurrence (Ballance et al. 2006).

In terms of POC fluxes, CCFZ belongs to the broadly defined mesotrophic abyss of Hannides and Smith (2003). As noted by Smith (2013), the east-west and south-north gradients in surface primary productivity are reflected in the flux of organic matter to the abyss. The flux rate were reported to vary temporally, the downward transport being particularly intensive in spring (Yamazaki and Kajitani 1999). That sedimentation may result in deposition of the phytodetritus on the sediment surface, particularly in depressions of the surficial layer (Smith et al. 1996, Radziejewska 2002, Thiel et al. 1988/1989, Radziejewska, pers. observ.) Smith et al. (1996) measured the plant pigment contents [chloroplastic pigment equivalent (CPE) = sum of chlorophyll a and phaeopigment contents; e.g. Neira et al. 2001] in sediments of the abyssal seafloor in the sub- and equatorial Pacific (at depths of about 4,000–5,000 m) on a transect running along 140°W from 12°S do 9°N, and reported contents averaging 0.36–9.8 µg/g dry sediment. Fluorometric assays made on the sediment samples collected in the eastern part of CCFZ in April–May 1997, after patches of phytodetritus had been observed on the core sediment surface, showed CPEs reaching 4.75 µg/g dry sediment (Radziejewska 2002).

The near-bottom currents were observed to vary greatly in direction and speed, the variations reflecting surface conditions produced by, i.a. wind-driven processes (hurricanes), and local modifications resulting from the seafloor relief (Tkatchenko and Stoyanova 1998). The CCFZ water dynamics shows periodic occurrence of wind-driven surface gyres the energy of which is transferred, as benthic storms, down to the near-bottom water (Kontar and Sokov 1994; Nikitin 1997; Sokov and Demidova 1992). Overall, Morgan et al. (1999) point out to the presence of 3 dynamic regimes in CCFZ: calm periods, with minimum current speeds and low tidal activity; intermediate, with mostly tidal periods, characterized by alteration of current speed (0 to 5–6 cm/s) and direction; and benthic storms, with current speeds increasing to 24-h averages of 8 cm/s. Effects of intensified dynamics of the near-bottom current were visible as levelling off and erosion of anthropogeni-cally disturbed sediment surface (Tkatchenko and Radziejewska 1998): 22 months after the sediment surface was altered by about 5–10 cm deep grooves flanked by mounds of overturned sediment, the grooves appeared to have been partially filled by the sediment and mound contours smoothed out (cf. Fig. 4.5 in Chap. 4).

The CCFZ seafloor is slightly inclined westwards, as evidenced by average depths increasing from about 4,000 m in the eastern part to about 5,400 m in the western part. The abyssal bottom relief shows a number of linearly arranged depressions, terraces, and elevations. Characteristic of the seafloor in the area is the presence of topographic structures that are volcanic in origin (IOM 1993; Kotliński 1998b). Active volcanism and /or hydrothermal venting is detectable in the patterns of physical and chemical characteristics of water masses overlying the CCFZ bottom, e.g. metal contents (Tkatchenko et al. 1997). Generally, however, the area is characterised by the depositional type of seafloor relief; the slope angle does not exceed 3°, the elevation difference averaging 100–200 m (IOM 1993; Kotliński 1998b). The sedimentary cover is relatively thin: it varies in thickness from about 50 to about 200 m. The surface layer is formed by recent sediments in the form of semi-liquid siliceous ooze, with up to 0.56 % organic carbon content (IOM 1993), and is strongly bioturbated (Pope et al. 1996; Radziejewska, pers. obs.).

As already pointed out above, the stretch of the abyssal plain seafloor flanked by the Clarion and Clipperton Fractures is a gigantic field of mineral deposits known as polymetallic nodules (**x. 2**). A characteristics of the field, including the description of the origin and the assessment of the nodule resources was provided by Kotlinski (1998b, 1999), Andreev and Gramberg (1998), and Morgan (2000). It has to be pointed out, however, that research aimed at detailed assessment of resources of nodules and metals and minerals they contain is in progress and is a main focus of contracts for exploration concluded by individual contractors with ISA (Kotliński, pers. comm.). Polymetallic nodules that cover the bottom at estimated abundances frequently averaging about 15 kg/m² and occasionally reaching 75 kg/m² lie on the surface of the semi-liquid sediment and may be partly embedded in it; some nodules are found buried in the sediment (Andreev and Gramberg 1998; Hein and Petersen 2013; Iljin et al. 1994, 1997; Kotliński 1998b). The conservative estimate of nodule amount in CCFZ, as given by Hein and Petersen (2013) is 21,100 million dry metric tonnes. Vast stretches of nodule deposits on the seafloor are interspersed by larger or smaller nodule-free areas (cf. Fig. 2.2). As shown by geotechnical assays, properties (mean grain size, water content, shear stress, plasticity) of the sediment in nodule-free areas differ from those shown by the nodule-bearing sediment (Radziejewska and Modlitba 1999).

In the uppermost 5 cm layer, the nodule-bearing sediment was found to be coarser (grains > 0.063 mm in diameter accounted for 39 % as opposed to 11.4 % in the nodule-free area), less hydrated, and less plastic than the nodule-free sediment, the nodule-bearing sediment shear stress increasing sharply downcore.

The environmental variability in the abyss, although known to be much more extensive than previously thought, is still far below the variability that can be induced by human activities. How resilient are abyssal communities to such variability remains unknown, although they have been presumed (and perceived) to be relatively poorly resilient (Davies et al. 2007; Ramirez-Llodra et al. 2011). And yet, as already indicated, plans are underway with which to effect the anthropogenic intervention into the deep sea (Lodge et al. 2014; Van Dover et al. 2014). It is therefore advisable to look for proxies which would provide an aid in the assessment of the degree of resilience, including the biotic community recovery potential. The next chapter will introduce the meiobenthos, a category of seafloor-dwelling organisms which—as already indicated in the previous chapter—has a potential of being used as such a proxy.

References

Alldredge AL, Silver MW (1988) Characteristics, dynamics and significance of marine snow. Prog Oceanog 20:41–82

Aller JY (1997) Benthic community response to temporal and spatial gradients in physical disturbance within a deep-sea western boundary region. Deep-Sea Res I 44:39–69

Amador JA, Alfaro EJ, Liyano OG et al (2006) Atmospheric forcing of the eastern tropical Pacific: a review. Prog Oceanog 69:101–142

Andreev SI, Gramberg IS (1998) The explanatory note to the metallogenic map of the world ocean. VNIIOkeangeologiya, St. Petersburg

Angel MV (1984) Detrital organic fluxes through pelagic ecosystems. In: Fasham MJ (ed) Energy and materials in marine ecosystems. Theory and practice. Plenum Press, London

Angel MV, Rice TL (1996) The ecology of the deep ocean and its relevance to global waste management. J Appl Ecol 33:915–926

Ballance LT, Pitman RL, Fiedler PC (2006) Oceanographic influences on seabirds and cetaceans of the eastern tropical Pacific: a review. Prog Oceanog 69:360–390

Barash MS, Kruglikova SB, Mukhina VV (2000) Stratigraficheskiye osobennosti osadochnykh obrazovaniy provintsii Klarion-Klipperton (Vostochnaya ekvatorialnaya Patsifika). Okeanologia 40:424–433

Billett DSM, Lampitt RS, Rice AL et al (1983) Seasonal sedimentation of phytoplankton to the deep sea benthos. Nature 302:520–522

Brown J, Colling A, Park D et al (1989) Ocean chemistry and deep sea sediments. The Open University, Milton Keynes

Coxall HK, Wilson PA, Pälike H et al (2005) Rapid stepwise onset of Antarctic glaciation and deeper calcite compensation in the Pacific Ocean. Nature 433:53–57

Davies AJ, Roberts JM, Hall-Spencer J (2007) Preserving deep-sea natural heritage: emerging issues in offshore conservation and management. Biol Conserv 138:299–312

Demidova T (1999) The physical environment in nodule provinces of the deep sea. In: Deep-seabed polymetallic nodule exploration: development of environmental guidelines. Proceedings of International Seabed Authority's workshop held in Sanya, Hainan Island, People's Republic of China, 1–5 June 1998. International Seabed Authority, Kingston, Jamaica

Fang J, Barcelona MJ, Nogi Y et al (2000) Biochemical implications and geochemical significance of novel phospholipids of the extremely barophilic bacteria from the Marianas Trench
at 11,000 m. Deep-Sea Res I 47:1173–1182

Fiedler PC, Talley LD (2006) Hydrography of the eastern tropical Pacific: a review. Prog
Oceanog 69:143–180

Fileman TW, Pond DW, Barlow RG et al (1998) Vertical profiles of pigments, fatty acids and
amino aids: evidence for undegraded diatomaceous material sedimenting to the deep ocean in
the Bellingshausen Sea, Antarctica. Deep-Sea Res I 45:333–346

Fowler SW, Knauer GA (1986) Role of large particles in the transport of elements and organic
compounds through the oceanic water column. Prog Oceanog 16:147–194

Gage JD, Tyler PA (1985) Growth and reproduction of the deep sea urchin Echinus affinis. Mar
Biol 90:41–53

Gage JD, Tyler PA (1991) Deep-sea biology: a natural history of organisms at the deep-sea floor.
Cambridge University Press, Cambridge

Gooday AJ (1994) The biology of deep-sea foraminifera: a review of some advances and their
applications in paleoceanography. Palaios 9:14–31

Hammond DE, McManus J, Berelson WM et al (1996) Early diagenesis of organic material in
Equatorial Pacific sediments: stoichiometry and kinetics. Deep-Sea Res 43:1365–1380

Hannides AK, Smith CR (2003) The northeastern Pacific abyssal plain. In: Black KD, Shimmield
GB (eds) Biogeochemistry of marine systems. Blackwell, Oxford

Hein JR, Petersen S (2013) The geology of manganese nodules. In: Baker E, Beaudoin Y (eds)
Deep sea minerals: manganese nodules, a physical, biological, environmental, and technical
review, vol 1B. Secretariat of the Pacific Community, GRID-Arendal

Hochachka PW (ed) (1976) Biochemistry at depth. Pressure effects on biochemical systems of
abyssal and midwater organisms: the 1973 Kona expedition of the "Alpha Helix". Pergamon
Press, Oxford

Hyun J-H, Kim K-H, Jung H-S et al (1998) Potential environmental impact of deep-seabed manganese nodule mining on the Synechococcus (Cyanobacteria) in the Northeast Equatorial
Pacific: effect of bottom water-sediment slurry. Mar Geores Geotechnol 16:133–143

Iljin AV, Ivakin AN, Lysanov YP (1994) O raspredelenii zhelezomargantsevykh konkreciy na poligone Klarion-Klipperton. Okeanologia 34:911–914

Iljin AV, Bogorov GB, Skorniakova NS (1997) O prostranstwennoy izmenchivosti zalegan-
iya zhelezo-margantsevykh konkretsiu (na poligone Klarion-Klipperton). Okeanologiya
37:285–294

IOM (1993) Geologiya, konkrecyonosnost' i prirodnye usloviya raiona pervonachalnoy
deyatel'nosti SO Interokeanmetall. IOM, Szczecin

Johnson GC, Toole JM (1993) Flow of deep and bottom waters in the Pacific at 10° N. Deep-Sea
Res I 40:371–394

Kalogeropoulou V, Bett BJ, Gooday AJ et al (2010) Temporal changes (1989–1999) in deep-sea
metazoan meiofaunal assemblages on the Porcupine Abyssal Plain, NE Atlantic. Deep Sea
Res II 57:1383–1395

Khripounoff A, Vangriesheim A, Crassous P (1998) Vertical and temporal variations of particle
fluxes in the deep tropical Atlantic. Deep-Sea Res I 45:293–316

Kolosov VP (1988) Chlorofill "a" i pervichnaya produkciya. In: Simonov AI (ed)
Ekologicheskiye usloviya vostochno-ekvatoriyalnoy oblasti severnoy chasti tikhogo okeana.
Gidrometeoizdat, Moskva

Kontar EA, Sokov AV (1994) A benthic storm in northeastern tropical Pacific over the fields of
manganese nodules. Deep-Sea Res 41:1069–1089

Kotliński R (1998a) Konkrecje polimetaliczne. In: Depowski S, Kotliński R, Rühle E et al (eds)
Surowce mineralne mórz i oceanów. Wydawnictwo Naukowe Scholar, Warszawa

Kotlinski R (1998b) The present state of knowledge on oceanic polymetallic ores as exemplified
by Interoceanmetal Joint Organization's activity. Mineral Pol 29:77–89

Kotliński R (1999) Metallogenesis of the world's ocean against the background of oceanic crust
evolution. Special Paper 4, Polish Geological Institute

Kotliński R, Rühle E (1998) Geneza i geologia oceanów. In: Depowski S, Kotliński R, Rühle E et al (eds) Surowce mineralne mórz i oceanów. Wydawnictwo Naukowe Scholar, Warszawa

Kotliński R, Stoyanova V, Tkatchenko G (1996) Environmental studies on a reference transect in the IOM pioneer area. In: Chung JS, Das BM, Roesset J (eds) Proceedings of 6th ISOPE Conference, vol 1. Los Angeles, USA, pp 54–57

Lodge M, Johnson D, Le Gurun G et al (2014) Seabed mining: International Seabed Authority environmental management plan for the Clarion-Clipperton Zone. A partnership approach. Mar Policy 49:66–72

Mantyla AW, Reid JL (1983) Abyssal characteristics of the World Ocean waters. Deep-Sea Res 30:805–833

Menzies RJ (1965) Conditions for the existence of life on the abyssal sea floor. Oceanog Mar Biol Ann Rev 3:195–210

Mestas-Nunez AM, Miller AJ (2006) Interdecadal variability and climate change in the eastern trophical Pacific: a review. Prog Oceanog 69:267–284

Morgan CL (2000) Resource estimates of the Clarion-Clipperton manganese nodule deposits. In: Cronan DS (ed) Handbook of marine mineral deposits. CRC Press, Boca Raton

Morgan CL, Odunton NA, Jones AF (1999) Synthesis of environmental impacts of deep seabed mining. Mar Geores Geotechnol 17:307–356

Neira C, Sellanes J, Levin LA et al (2001) Meiofaunal distributions on the Peru margin: relationship to oxygen and organic matter availability. Deep-Sea Res I 48:2453–2472

Nienstedt JC, Arnold AJ (1988) The distribution of benthic foraminifera on seamounts near the East Pacific Rise. J Foram Res 18:237–249

Nikitin OP (1997) Vertikalnaya struktura sinopticheskikh techeniy v severo-vostochnom tropikal'nom Tikhom okeane. Okeanologia 37:819–831

Ozturgut E, Anderson GD, Burns RE et al (1978) Deep ocean mining of manganese nodules in the North Pacific: pre-mining environmental conditions and anticipated mining effects. NOÀA Techn Mem, ERL MESA-33

Pennington TJ, Mahoney KL, Kuwahara VS et al (2006) Primary production in the eastern tropical Pacific: a review. Prog Oceanog 69:285–317

Pfannkuche O, Boetius A, Lochte K et al (1999) Responses of deep-sea benthos to sedimentation patterns in the North-East Atlantic in 1992. Deep-Sea Res I 46:573–596

Pope RH, DeMaster DJ, Smith CR et al (1996) Rapid bioturbation in equatorial Pacific sediments: evidence from excess ^{234}Th measurements. Deep-Sea Res II 43:1339–1364

Rabalais NN, Diaz RJ, Levin LA et al (2010) Dynamics and distribution of natural and human-caused hypoxia. Biogeosci 7:585–619

Radziejewska T (2002) Responses of deep-sea meiobenthic communities to sediment disturbance simulating effects of polymetallic nodule mining. Int Rev Hydrobiol 87:459–479

Radziejewska T, Modlitba I (1999) Vertical distribution of meiobenthos in relation to geotechnical properties of deep-sea sediment in the IOM pioneer area (Clarion-Clipperton Fracture Zone, NE Pacific). In: Chung JS, Sharma R (eds), Proc 3rd Ocean Mining Symposium, Goa, India

Ramirez-Llodra E, Brandt A, Danovaro R et al (2010) Deep, diverse and definitely different: unique attributes of the world's largest ecosystem. Biogeosci 7:2851–2899

Ramirez-Llodra E, Tyler PA, Baker MC et al (2011) Man and the last great wilderness: human impact on the deep sea. PLoS ONE 6:e22588

Scharek R, Tupas LM, Karl DM (1999) Diatom fluxes to the deep sea in the oligotrophic North Pacific gyre at Station ALOHA. Mar Ecol Prog Ser 182:55–67

Schlitzer R (2004) Ocean Data View. http://odv.awi-bremerhaven.de

Seibold E, Berger WH (1993) The sea floor. An introduction to marine geology, 2nd edn. Springer, Berlin

Semina HI, Mikaelyan AS, Belyaeva GA (1997) Fitoplankton raznykh razmernykh grupp v severnoy subtropicheskoy zone Tikhogo okeana. Okeanologia 37:730–738

Smith CR (2013) Biology associated with manganese nodules. In: Baker E, Beaudoin Y (eds) Deep sea minerals: manganese nodules, a physical, biological, environmental, and technical review, vol 1B. Secretariat of the Pacific Community, GRID-Arendal

Smith CR, Baco AR (2003) The ecology of whale falls at the seep-sea floor. Oceanog Mar Biol Ann Rev 41:311–354

Smith CR, Hoover DJ, Doan SE et al (1996) Phytodetritus at the abyssal seafloor across 10° of latitude in the central equatorial Pacific. Deep-Sea Res II 43:1309–1338

Smith CR, Berelson W, DeMaster DJ et al (1997) Latitudinal variations in benthic processes in the abyssal equatorial Pacific: control by biogenic particle flux. Deep-Sea Res II 44:2295–2317

Smith CR, Maybaum HL, Baco AR et al (1998) Sediment community structure around a whale skeleton in the deep Northeast Pacific: macrofaunal, microbial and bioturbation effects. Deep-Sea Res II 45:335–364

Snelgrove RVP, Butman CA (1994) Animal-sediment relationships revisited: cause versus effect. Oceanog Mar Biol Ann Rev 32:111–177

Sokov AV, Demidova TA (1992) Ob antitsiklonicheskom vikhre v severo-vostochnoy chasti tropicheskoy zony Tikhogo okeana. Meteorolog Gidrolog 3:57–64

Stoecker DK, Gustafson DE, Verity PG (1996) Micro- and mesoprotozooplankton at 140°W in the equatorial Pacific: heterotrophs and mixotrophs. Aquat Microb Ecol 10:273–282

Thiel H, Pfannkuche O, Schriever G et al (1988/1989) Phytodetritus on the deep-sea floor in a central oceanic region of the northeast Atlantic. Biol Oceanog 6:203–239

Thistle D (1998) Harpacticoid copepod diversity at two physically reworked sites in the deep sea. Deep-Sea Res II 45:13–24

Thistle D, Levin LA (1998) The effect of experimentally increased near-bottom flow on metazoan meiofauna at a deep-sea site, with comparison data on macrofauna. Deep-Sea Res I 45:625–638

Thistle D, Levin LA, Gooday AJ et al (1999) Physical reworking by near-bottom flow alters the metazoan meiofauna of Fieberling Guyot (northeast Pacific). Deep-Sea Res I 46:2041–2052

Tholosan O, Garcin J, Bianchi A (1999) Effects of hydrostatic pressure on microbial activity through a 2000 deep water column in the NW Mediterranean Sea. Mar Ecol Prog Ser 183:49–57

Tivey MK, Bradley AM, Joyce TM et al (2002) Insights into tide-related variability at seafloor hydrothermal vents from time-series temperature measurements. Earth Planet Sci Lett 202:693–707

Tkatchenko GG, Radziejewska T (1998) Recovery and recolonization processes in the area disturbed by a polymetallic nodule collector simulator. In: Chung JS, Olagnon M, Kim CH et al (eds) Proceedings of 8th ISOPE Conference, vol 2. Montreal, Canada, pp 282–286

Tkatchenko G, Stoyanova V (1998) The scale and nature of water column variability in an area designated for polymetallic nodule mining. In: Chung JS, Olagnon M, Kim CH et al (eds) Proceedings of 8th ISOPE Conference, vol 1. Montreal, Canada, pp 39–43

Tkatchenko G, Kotlinski R, Stoyanova V et al (1997) On the role of geologic factors in determining peculiarities of water mass structure in the Clarion-Clipperton ore field. In: Chung JS, Das BM, Matsui T et al (eds) Proceedings of 7th ISOPE Conference, vol 1. Honolulu, Hawaii, pp 959–961

Tsuchiya M, Talley LD (1996) Water property distributions along an eastern Pacific hydrographic section at 135°W. J Mar Res 54:541–564

Tyler PA (1988) Seasonality in the deep sea. Oceanog Mar Biol Ann Rev 26:227–258

Tyler PA (1995) Conditions for the existence of life at the deep-sea floor: an update. Oceanog Mar Biol Ann Rev 33:221–244

Van Dover CL, Aronson J, Pendleton L et al (2014) Ecological restoration in the deep sea: Desiderata. Mar Pol 44:98–106

Wang C, Fiedler PC (2006) ENSO variability and the eastern tropical Pacific: a review. Prog Oceanog 69:239–266

Whitmarsh RA, Bull JM, Rothwell RG et al (1996) The evolution and structure of ocean basin. In: Sommerhayes CP, Thorpe SA (eds) Oceanography. An Illustrated Guide. Manson Publishing, London

Wijffels SE, Toole JM, Bryden HL et al (1996) The water masses and circulation at 10°N in the Pacific. Deep-Sea Res I 43:501–544

Willett CS, Leben RR, Lavin MF (2006) Eddies and tropical instability waves in the eastern tropical Pacific: a review. Prog Oceanog 69:281–283

Wishner KF, Ashjian CJ, Gelfman C et al (1995) Pelagic and benthic ecology of the lower interface of the eastern tropical Pacific oxygen minimum zone. Deep-Sea Res I 42:93–115

Yamazaki T, Kajitani J (1999) Deep-sea environment and Impact Experiment to It. In: Chung JS, Matsui T, Koterayama W (eds) Proceedings of 9th ISOPE Conference, vol 1. Brest, France, pp 374–381

Young CM, Tyler PA (1993) Embryos of the deep sea echinoid *Echinus affinis* require high pressure for development. Limnol Oceanog 38:178–181

Chapter 3
Meiobenthos of the Sub-equatorial North-Eastern Pacific Abyssal Seafloor: A Synopsis

Abstract The metazoan meiobenthos of the sub-equatorial North-Eastern Pacific abyss, CCFZ included, groups a fairly high number of higher taxa (including nematodes, harpacticoid copepods, ostracods, kinorhynchs, tardigrades, gastrotrichs, halacaroid mites, loriciferans, meiobenthos-sized polychaetes), the free-living nematodes and harpacticoid copepods being dominant. Both groups are very diverse in terms of taxon (genus, species) richness, the genus-level lists from the studies conducted so far containing 10–246 and 34–62 genera of nematodes and harpacticoids, respectively. Most genera (e.g. *Acantholaimus* among the nematodes and *Pontostratoites* among the harpacticoids) seem to occur throughout CCFZ, although the dominant genera change depending on the location (e.g. the *Terschellingia* nematodes were found to dominate in the eastern part of CCFZ and *Acantholaimus* in most other locations sampled). Most of the individuals found in samples represent as yet unknown, undescribed species. Meiobenthic abundances were found to vary over a range on the order of 10^1–10^2 ind./10 cm^2, i.e. an order or two lower than the densities recorded in shallower and shelf waters. The variability in the abundance is thought to results from sediment- and habitat type-dependence (nodule-bearing *versus* nodule-free bottom, the nodules supporting characteristic faunas of their own) as well as from small-scale patchiness induced by natural factors (near-bottom water dynamics, activity of megafauna, presence of biogenic structures such as large protozoans the Xenophyophorea and the Komokiacea). The CCFZ meiobenthos was found to respond to episodic inputs of organic material in the form of phytodetritus sedimentation, some nematode (desmoscolecids) and harpacticoid (argestids) taxa being particularly responsive and increasing their abundance.

Keywords Meiobenthos · Deep sea · Pacific · CCFZ · Nematoda · Harpacticoida · Phytodetritus · Variability · Polymetallic nodules

© The Author(s) 2014   29
T. Radziejewska, *Meiobenthos in the Sub-equatorial Pacific Abyss*,
SpringerBriefs in Earth System Sciences, DOI 10.1007/978-3-642-41458-9_3

3.1 General Considerations

The origins of research on deep-sea benthos date back to the great oceanic expeditions of the 19th century. It was then that sediment-dwelling organisms were found throughout the world ocean's depth range, right down to the bottom of the oceanic trenches (Belayev 1989; Gage and Tyler 1991). For a long time, studies on the deep-sea benthos were limited primarily to the collection of specimens and their identification; that was, to quote Tyler (2003) a "heroic era" in the deep-sea research. Those studies resulted in the discovery and description of an immense number of new species and higher taxa. In consequence, the body of knowledge on the living world of the deep seafloor and oceanic biodiversity (understood as species or taxon richness) was greatly augmented, thereby laying a groundwork for the contemporary assessment of and discussion on biological diversity and conservation of the deep-sea part of the biosphere (Angel 1996; Angel and Rice 1996; Boucher and Lambshead 1995; Etter and Grassle 1992; Gray 1994; Gray et al. 1997; Rex and Etter 2010; Rex et al. 1993). Particularly noteworthy in this respect is the enhancement of data collection and interpretation resulting from the technological progress in seafloor observation. Implementation of photo and video surveys of the deep-sea bottom have resulted in a significant increase in the amount of data on the megafauna (=organisms, usually larger than 4 cm, observable and identifiable in underwater imagery; Bluhm 1994; Radziejewska and Stoyanova 2000; Tilot 1992), thus making it possible to observe, and in many cases to describe and identify, organisms never touched by the human hand (Belayev 1989).

While the megabenthos can be directly observed and quantified, research on the smaller components of the benthos (macro-, meio-, and microbenthos) requires an indirect approach via collection of sediment samples, necessary for both qualitative information and quantitative data. Qualitative information on the Pacific deep-sea meiofauna, i.e. information on the occurrence and taxonomic composition of the abyssal and hadal meiobenthic communities has been accumulating almost since the Pacific deep seafloor began to be scientifically explored (Thiel 1983; Belayev 1989). However, not always were the meiobenthic organisms included in the early phase of the deep-sea benthos research in the Pacific. An eminent example of disregard for the abyssal meiobenthos is provided by one of the major earlier publications on Pacific biology—a book edited by Zenkevich (1969) which—despite a remarkable richness of faunistic information—lacks any reference to the benthic meiofauna.

As elsewhere in the marine science, purely qualitative studies on the deep-sea benthos, although still pursued, were with time complemented by the quantitative approach, prevalent in the present era of exploration (Tyler 2003). Efforts have been, and are being, made to assess the abundance, biomass, metabolic activity indices, production, and diversity of deep-sea benthic communities, as such data are urgently needed for resolving a number of questions regarding phenomena that proceed on different temporal and spatial scales. Quantitative information is amenable to hypothesis testing and makes it possible to draw inferences on relationships between community parameters and biotic and abiotic environmental constraints. This is particularly relevant when effects of human intervention into the deep sea are to be assessed, as those effects will have to be prised apart from the naturally occurring variability the range of which is in most instances still very poorly known.

In this context it is necessary to emphasise that, despite the progress already made, the knowledge on deep-sea benthic ecology, particularly with respect to the smallest benthos compartment (meiofauna and microorganisms), is still far from adequate, and it still takes a great effort to broaden it. The need for such effort is strengthened by achievements of the large-scale international research programmes such as the Joint Global Ocean Flux Study (JGOFS; Fasham et al. 2001), and has been emphasised by results of the Census of Marine Life programme (Costello et al. 2010). As already mentioned, the insights gained through those programmes have led to modification of the time-honoured views on the natural variability (or the lack thereof) of environmental conditions in the oceans' abyss. Records of benthic storms in the benthic boundary layer and their effects on the deep-sea floor (Aller 1997; Kontar and Sokov 1994; Sokov and Demidova 1992), or of episodic phytodetritus sedimentation down to the abyssal seafloor during which the plant material reaches the abyssal sediment within about 3–6 weeks after the surface bloom (Drazen et al. 1998; Scharek et al. 1999) and triggers numerous chemical and biological processes on the bottom (Fileman et al. 1998; Gehlen et al. 1997; Turner 2002) have led to revision of the deeply rooted conviction that the deep-sea environment is extremely stable (Loubere 1998; Tyler 1988, 1995). However, the extent of natural variability in biological communities of the Pacific's abyssal seafloor still remains a great unknown.

Since their discovery in the late 1970s, a remarkable progress has been made in the research on deep-sea hydrothermal vents and their unique communities, the very existence of which relies on sulphide metabolism of chemoautrotrophs that supply energy for heterotrophic production of vent ecosystems (Hessler and Kaharl 1995). The growing body of information on the vent communities includes also some data on the meiobenthos (Dinet et al. 1988; Vanreusel et al. 1997; Zekely et al. 2006).

Most of the information on the deep-sea benthos, and meiobenthos in particular, collected to date originates from the Atlantic (e.g. Dinet et al. 1985; Flach et al. 1999; Galéron et al. 2001; Gooday et al. 1996; Soltwedel 1997, 2000; Tietjen et al. 1989; Vincx et al. 1994). Since the early 1990s, however, interest in the Pacific deep-sea meiobenthos has been observed to grow dynamically as well (e.g. Alongi 1992; Hannides and Smith 2003; Lambshead et al. 2002; Shirayama and Kojima 1994). That interest was doubtless spurred by questions arising from studies on communities associated with hydrothermal venting (Vanreusel et al. 1997) and problems related to the perceived and projected global change and its effects on living communities and biological diversity patterns (Hannides and Smith 2003; Rex et al. 2000). Moreover, questions posed by future anthropogenic activities in the deep sea, such as waste storage, CO_2 sequestration, and the anticipated development of mineral resources (Renaud-Mornant and Gourbault 1990; Thiel 1991, 1992; Thistle et al. 2007b) provided an additional stimulus for studies on the abyssal meiofauna in the Pacific.

3.2 Methodological Problems

To obtain a reliable set of quantitative data on a meiobenthic community, it is necessary to collect appropriate sediment samples, i.e. to use a collecting gear that will retrieve and bring on board samples of intact (undisturbed) sediment overlain with a transparent layer of the near-bottom water. In contemporary research, the desired quality of sediment samples is enhanced by the use of the multiple corer of Barnett et al. (1984) (commonly referred to as the *multicorer*) (Bett et al. 1994). Before the multiple corer came into common use in the deep-sea research, the sediment for quantitative meiobenthos studies had been collected by subsampling the contents of various grabs and box corers (Shirayama 1984a, b; Shirayama and Kojima 1994; Soltwedel 2000; Thiel 1983). Among the latter, the most popular gear is the so-called USNEL-type box corer (Gage and Tyler 1991), used most widely in the deep-sea benthos research; it retrieves a 50 × 50 cm surface area sediment monolith. For meiobenthic studies, the sediment in the box corer was subsampled with various small-size devices such as, e.g. the *Meiostecher* of Thiel (1983), a vegematic (Trueblood et al. 1997) or plastic cylinders of various diameter (e.g. Alongi 1992; Renaud-Mornant and Gourbault 1990; Shirayama 1984a, b; Snider et al. 1984).

Detailed tests reported on by Bett et al. (1994) showed, however, that—as opposed to the multicorer—a box corer cannot be regarded as a gear fully suitable for obtaining reliable quantitative meiobenthos samples as the surficial semi-fluid layer of the sediment contained in the box-corer barrel is usually disturbed by the so-called bow-wave the gear produces on impact with the bottom. An additional disturbance may be introduced during the corer retrieval and handling on board. As the surficial layer normally holds the bulk of the meiobenthos, preservation of the original condition and form of that layer is crucial for collecting reliable quantitative data. As demonstrated by Bett et al. (1994) and by Shirayama and Fukushima (1995), the loss of and /or disturbance to the sediment surface layer is eliminated when the multiple corer is used. In addition, a sediment core collected with the multicorer can be reliably sliced, i.e. divided into layers, thus making it possible to follow patterns of vertical distribution of the meiobenthos in sediment and to observe and record changes in abiotic parameters, lithology and geochemistry of abyssal sediments and their pore water along a downcore profile (Barnett et al. 1984). Because of a relatively small surface area of the multicorer's tubes (usually 9–10 cm), the entire 0.5–3 cm thick slices can be processed, although some workers (e.g. Kaneko (Sato) et al. 1995, 1997) chose to subsample the multicorer-collected sediment. However, when subsampling a small diameter deep-sea sediment core a risk of missing representatives of rare taxa is increased (Galtsova and Kamenskaya 1993).

3.3 Composition of the Abyssal Pacific Meiofauna

A number of studies on the deep-sea meiobenthos (e.g. Alongi 1992; Gooday 2002; Hessler and Jumars 1974; Jumars and Hessler 1976; Snider et al. 1984; Soltwedel et al. 2003; Radziejewska, unpublished data) have demonstrated the prominence of meiofaunal-sized protozoans, the Rhizaria (Foraminifera), in the deep sea sediments. Although those protozoans are extremely interesting from many points of view, e.g. taxonomic (Gooday et al. 2005), genetic and phylogenetic (Pawlowski and Holzman 2008), and also because they are known to respond very rapidly—by increasing their abundance—to a pulse of phytodetritus sedimentation to the abyssal seafloor (e.g. Gooday et al. 1996; Thiel 1983), most publications on the deep-sea meiobenthos focus primarily on the metazoan component of the meiofauna. This approach is adopted in this work as well.

Reports on the Pacific abyssal meiobenthos list the major taxa that are certain to occur in all the sediment samples. The group of such taxa includes primarily the free-living nematodes (Nematoda) and small benthic copepods (Copepoda Harpacticoida). Fairly regularly will one encounter turbellarians (Turbellaria), cyclopoid copepods (Copepoda Cyclopoida), ostracods (Ostracoda), kinorhynchs (Kinorhyncha), tardigrades (Tardigrada), and meiobenthos-sized polychaetes (Polychaeta). Less frequent are gastrotrichs (Gastrotricha), priapulids (Priapulida), halacaroid mites (Halacaroidea), and loriciferans (Loricifera), a phylum described in Kristensen (1983). All those taxa belong to the so-called *permanent meiobenthos* (Giere 2009; McIntyre 1969), i.e. meiobenthic organisms that remain within the meiobenthic size class throughout their respective life cycles and retain meiobenthic biological characteristics (e.g. lack of pelagic larvae, low number of eggs, short generation time). Deep-sea sediment samples contain also representatives of the so-called *temporary meiobenthos* (Giere 2009; McIntyre 1969) comprising larval and juvenile stages of the macrobenthos, such as polychaetes (Polychaeta), cumaceans (Cumacea), tanaidaceans (Tanaidacea), isopods (Isopoda), and amphipods (Amphipoda) among crustaceans as well as gastropods (Gastropoda) and bivalves (Bivalvia) among molluscs. It needs to be borne in mind, however, that—due to the tendency for miniaturisation observed in the deep-sea organisms (Thiel 1975), particularly among taxa that do not usually grow to a large size (e.g. McClain et al. 2005)—the deep-sea meiobenthic size class range is narrower, with 0.250 mm being regarded as the upper cut-off point (Thiel 1975). Therefore, representatives of certain taxa regarded as typically meiobenthic in the shallow water studies may feature prominently within the deep-sea macrobenthos, e.g. some larger nematodes (Sharma and Bluhm 2010).

The description below will focus on the two metazoan meiobenthic taxa which are quantitatively most important in the abyss: nematodes and harpacticoid copepods.

Nematoda. Similarly to the seafloor in shallower localities, nematodes are the commonest and most abundant meiobenthic organisms in all the Pacific's abyssal and hadal areas investigated. They usually contribute more than 90 % of all the metazoans found in samples (Gage and Tyler 1991; Hannides and Smith 2003). The specific status of nematodes in the abyssal meiobenthos is emphasised by their high diversity, manifested through high species, genus, and higher taxon richness (Table 3.1).

The data presented in the table are difficult to compare directly, as some authors (e.g. Lambshead et al. 2002; Renaud-Mornant and Gourbalt 1990) reported estimates of species richness (number of species) per sample or station [ES(51) and Margaleff's index, respectively], whereas others provided data from direct identification analyses. It should be added that most meiobenthic species, and very probably many genera as well, found at the abyssal depths have not been described; the majority are new to science and, for practical purposes, are usually denoted with a symbol in published taxonomic lists as the so-called putative species (e.g. Vopel and Thiel 2001).

As can be seen in Table 3.1, most of the data sets were derived from the Clarion-Clipperton area of the central-sub-equatorial part of the Pacific (CCFZ), with a single set of data from another eastern Pacific abyssal nodule field, the Peru Basin, referred to here for comparative purposes. Estimates of taxonomic richness at the genus level for CCFZ are seen to differ widely between studies. While Renaud-Mornant and Gourbault (1990) identified as few as 45 nematode species representing several genera (10 dominant; see below), at the other end of the range are Wilson and Hessler (1987 quoted by Bussau et al. 1995) with 145 genera; re-analysis of data originally processed by Radziejewska et al. (2001), including validation check against the World Register of Marine Species (WoRMS) database (www.marinespecies.org) in 2013–2014 yielded a total number of 246 valid genera; Miljutina et al. (2010) found the nematodes in their materials to represent 325 species spread among 97 genera. The differences between various studies deserve a closer look. They may have arisen from a multitude of reasons; on the one hand, the differences could be due to technical details, e.g. gear type and /or sample size (number of cores processed), while on the other, more fundamental processes could have been manifested here, such as differences between areas due to their geographical location (restricted connectivity and/or taxon turnover between the areas sampled) and prevalence of habitat types (e.g. nodule-bearing and nodule-free bottoms). Data sets deriving from different years could differ because of still another reason, the temporal variability. Although earlier thought to be unimportant in the deep sea, temporal differences do manifest themselves in the abyss (Copley et al. 2010; Glover et al. 2010; Kalogeropoulou et al. 2010). One example was furnished by the study of Radziejewska et al. (2001): while their 1995 set of samples collected from a nodule-free area yielded a total of 226 genera, the samples collected in 1997 from the same area, just prior to the onset of a strong El Niño Southern Oscillation (Kim et al. 2011, 2012), and following an episode of phytodetritus deposition on the seafloor were found to contain 171 genera.

Regardless of the differences between various studies, it may be concluded that the numbers of genera reported, particularly those in more recent studies, evidence a considerable genus-level (and apparently the species-level as well) diversity at the local scale (alpha diversity). This is also a conclusion stemming from the

Table 3.1 Pacific abyssal nematode taxon richness estimates (in chronological order of references)

Area	Depth range (m)	Reference	Gear (BC, box corer; MC, multicorer); no. of cores	No. of genera	No. of (putative) species	Remarks
Eastern-central Pacific	4,400–4,600	Wilson and Hessler (1987) in Bussau et al. (1995)	BC, no. of cores not reported by Bussau et al. (1995)	148	Not reported	
CCFZ, central part	4,960–5,154	Renaud-Mornant and Gourbault (1990)		About 10 reported	45	Species richness (estimated) per station
Peru Basin (south sub-equatorial Pacific)	4,037–4,158	Bussau and Lorenzen (1991), Bussau and Vopel (1999), Bussau et al. (1995), Vopel and Thiel (2001)	MC; 10 (Vopel and Thiel 2001)	68	About 300 (Bussau and Lorenzen 1991); not reported, but at least 97 new to science	
CCFZ, eastern part	4,380–4,430	Radziejewska et al. (2001a)	MC; 30 cores	246	Not reported	Validity checked against the World Register of Marine Species database (www.marinespecies.org)
Central equatorial Pacific (incl. CCFZ)	4,301–4,994	Lambshead et al. (2002)	BC, 15 cores; MC, 6 cores	Not reported	40–58	Morphospecies approach used; estimated [ES(51)] species richness per core reported
Central equatorial Pacific (incl. CCFZ)	4,301–4,994	Lambshead et al. (2003)	BC, 15 cores; MC, 6 cores	32 Dominant and subdominant (>5 % total number of individuals) genera reported only	200 Reported; 159–386 estimated	Area of Lambshead et al. (2002) paper; total number of morphospecies reported

(continued)

Table 3.1 (continued)

Area	Depth range (m)	Reference	Gear (BC, box corer; MC, multicorer); no. of cores	No. of genera	No. of (putative) species	Remarks
CCFZ, central part	4,947–5,046	Miljutina et al. (2010)	BC, 2 cores; MC, 21 cores	97	325	
CCFZ (central part)	4,860	Radziejewska et al. (unpublished)[a]	BC, 4 cores in 1993; MC, 3 cores in 1994	124	Not reported	

[a]Conference paper: Radziejewska T, Thistle D, Galtsova VV, Kulangieva L, Drzycimski I, Trueblood DD, Ozturgut E (2006) Benthic impact experiments in the Clarion-Clipperton Fracture Zone (equatorial NE Pacific): a comparison of meiofaunal responses to seafloor disturbance in two nodule-bearing areas. In: 11th Deep-Sea Biology Symposium, Southampton, 9–14 July 2006 (Book of Abstracts, p 70)

relatively recent paper of Miljutin et al. (2010) who, after a comprehensive global-scale deep-sea literature review, stated that there were now 638 nematode species known to science, belonging to 175 genera and 44 families. It is obvious from the review presented above that the number of valid genera present in the deep seafloor is much higher, but Miljutin et al. (2010) confined their analysis to those genera to which valid species could be ascribed.

In addition to differences in the number of genera recorded by various authors, the dominant genera and families are observed to differ as well (Table 3.2), there being also between-years differences in dominant genera. For example, the data for 1995 and 1997 analysed by Radziejewska et al. (2001) showed the 1995 set to be dominated by *Terschellingia, Halalaimus*, and *Linhystera*, while the 1997 set showed a pronounced change in the dominance structure, the domination being taken over by *Desmoscolex* and *Pareudesmoscolex* accompanied, as subdominants, by *Terschellingia, Linhystera, Paralongicyatholaimus*, and *Paracanthonchus*. The change of dominant genera was ascribed to the ability of small-sized nematodes (primarily the desmoscolecids *Desmoscolex* and *Pareudesmoscolex*) to exploit episodic bouts of carbon supply arriving with phytodetritus deposition (Vanaverbeke et al. 2004), as such an event was inferred to have happened immediately prior to the 1997 sampling (Radziejewska 2002; see below).

The data reported in Table 3.2 highlight two major points: (1) the degree of dominance is usually rather low (the evenness is high), i.e. there are usually numerous genera accounting for more than the assumed threshold proportion of the individuals examined, and the threshold is set at a low level, 10 % at the maximum; besides, usually none of those species listed as dominants or subdominants assumes an overwhelming dominance, as is often the case in shallow-water habitats (e.g. Steyaert et al. 1999); (2) comparison of all the data sets reveals that the dominance structure differs between the localities sampled in any single study (e.g. Lambshead et al. 2002, 2003), changes on the scale of the entire CCFZ (with *Terschellingia* dominating in the eastern part of CCFZ and *Acantholaimus* in other areas; Radziejewska et al. 2001; Vanreusel et al. 2010), and differs between the two major Pacific abyssal nodule provinces, the CCFZ and the Peru Basin [cf. data in Bussau and Lorenzen (1991); Bussau et al. (1995); Vopel and Thiel (2001)]. The causes underlying those differences can be at best speculated at, the putative drivers including (micro)habitat differences (nodule-bearing *versus* nodule-free bottom; e.g. Miljutina et al. 2010; Radziejewska and Modlitba 1999; Vanreusel et al. 2010), proximity or otherwise to hydrothermal vent areas of the East Pacific Rise (Vanreusel et al. 2010), larger- or abyss-scale oceanographic effects (Lambshead et al. 2002, 2003), and temporal effects. Since such differences affect our general perception of the abyssal diversity, they deserve more detailed studies.

As already mentioned, one should expect that most individuals found in the samples belong to as yet unknown, undescribed species (Boucher and Lambshead 1995). For example, while the Peru Basin samples processed by Bussau and Vopel (1999) contained at least 11 new genera and 97 new species (K. Vopel, personal communication), only 3 and 6 of them, respectively, were formally described (Bussau and Vopel 1999). All the nematode species reported by Lambshead et al.

Table 3.2 Nematode families and genera dominant in various studies

Area	Source	Dominance level adopted	Dominant families	Dominant genera	Subdominant genera
CCFZ, central part	Renaud-Mornant and Gourbault (1990)	Not reported	Ironidae, Desmodoridae	*Syringolaimus, Molgolaimus, Acantholaimus, Desmoscolex, Tricoma, Leptolaimus, Halalaimus*	Not reported
Peru Basin (south sub-equatorial Pacific)	Bussau and Lorenzen (1991), Bussau and Vopel (1999), Bussau et al. (1995), Vopel and Thiel (2001)	>10 %	Chromadoridae, Desmoscolecidae, Diplopeltidae, Microlaimidae, Oxystominidae, Xyalidae, and Monhysteridae	*Acantholaimus, Thalassomonhystera, Halalaimus,* unknown xyalid, *Desmoscolex, Diplopeltula*	Reported as "other abundantly represented genera": *Microlaimus, Paracyatholaimus, Molgolaimus, Camacolaimus, Syringolaimus, Trefusia, Tricoma*
Central equatorial Pacific (incl. CCFZ)	Lambshead et al. (2003)	>5 % For dominants; >1 % for subdominants	Monhysteridae, Chromadoridae, Microlaimidae, Xyalidae, Oxystominidae	*Thalassomonhystera, Acantholaimus, Molgolaimus, Aponema, Microlaimus, Diplopeltoides, Acantholaimus, Linhystera, Quadricoma, Halalaimus, Desmoscolex*	Order of dominant families and genera different between locations
CCFZ, eastern part	Radziejewska et al. (2001)	>1 %	Desmoscolecidae, Cyatholaimidae, Xyalidae, Linhomoeidae, Chromadoridae, Oxystominidae	*Terschellingia, Desmoscolex, Pareudesmoscolex, Halalaimus, Linhystera, Paralongicyatholaimus, Paracanthonchus*	Order of dominant families and genera different beween years
CCFZ, central part	Miljutina et al. (2010)	>4 %	Xyalidae, Monhysteridae, Chromadoridae, Desmoscolecidae, Oxystominidae	*Thalassomonhystera, Acantholaimus, Theristus, Desmoscolex, Marisalbinema*	Dominance order in families depending on the presence of nodules (Xyalidae in nodule-free and Monhysteridae in nodule bearing bottom)
CCFZ (central part)	Radziejewska et al. (unpublished)[a]	>1 %	Chromadoridae, Linhomoeidae, Cyatholaimidae, Micro-laimidae, Diplopeltidae	*Comesoma, Chromadora, Metalinhomoeus, Terschellingia, Longicyatholaimus, Bolbolaimus, Microlaimus, Descmoscolex*	Dominance in families and genera strongly dependent on the year/gear used

[a]see Footnote to Table 3.1

(2002, 2003) were considered to be new to science. Taxonomic identification and description of new species, particularly within meiobenthic taxa, is a very tedious task, undertaken by regrettably fewer and fewer taxonomists. For this reason, despite the assumed high species richness of the abyssal nematofauna, relatively few new deep-sea nematode species have so far been formally described from the Pacific. In his work on the hadal fauna, Belayev (1989) mentioned Timm's (1970 in Belayev 1989) descriptions of 4 new desmoscolecid nematode species (*Desmoscolex bathybius*, *D. gladisetosus*, *D. volifer*, and *Quadricoma desmoscolecoides*) from the Peru Trench. Kito (1983, in Decraemer and Gourbault 1997) provided description of two new *Cephalochaetosoma* species from off Philippines, found at 5,551 m. In addition to describing a new genus (*Dinetia* gen. n.), Decraemer and Gourbault (1997) identified and described a new species within that genus (*D. nycterobia*) and a new subspecies of *Cephalochaetosoma pacificum* (*C. pacificum notium*) collected from two different hydrothermal vent localities in the Pacific, but at depths (1,707–2,600 m) shallower than those formally regarded as abyssal. Decraemer and Gourbault (1997) commented on an unnamed genus and species of juvenile draconematid nematodes, collected from a polymetallic nodule field in the Peru Basin (4,990 m) and provisionally described by Bussau (1993) in his dissertation as *Eudraconema dracocephalum*; the species, however, was not amenable to a formal description due to the lack of appropriate type specimens. Verschelde et al. (1998), too, dealt with Pacific hydrothermal vent sediment nematodes and described 4 new species of desmodorid nematodes (*Desmodora alberti*, *D. marci*, *Desmodorella balteata*, and *Desmodorella spineacaudata*). The relatively recent descriptions include those by Bussau and Vopel (1999) who, as already mentioned, described 3 new genera and 6 new species of the family Microlaimidae. It seems that the taxonomic effort directed on nematodes has gained a new momentum in the recent years, as sampling effort in the CCFZ has been intensified (e.g. ISA 2007; Miljutina et al. 2010). This research has already brought description of new species; for example, Miljutin and Miljutina (2009a) described 3 new species of the family Microlaimidae. Moreover, in another recent paper, Miljutin and Miljutina (2009b) provided descriptions of new species representing the Benthimermithidae, a rare and little-known nematode family whose members, as juveniles, parasitise other invertebrates, while adults are free-living but do not feed (have no buccal cavity). It is thought that they derive nourishment exclusively via transepidermal absorption of dissolved organic matter (Galtsova 1991; Tchesunov 1997).

The research on deep-sea nematodes will be doubtless advanced by new approaches to taxonomy, such as DNA barcoding (Hebert and Gregory 2005; Lambshead 2011; but see Krishnamurthy and Francis 2012 for a critical view).

In addition to taxonomic identifications, the free-living marine nematodes are viewed also from the standpoint of their functional roles in the community. Those roles are inferred by using various morphological proxies (e.g., buccal cavity structure, tail shape; Schratzberger et al. 2007; Thistle et al. 1995; Fig. 3.1) thought to reflect certain functional aspects. Morphological examination of nematode individuals allows to assign them to the so-called trophic (feeding) types,

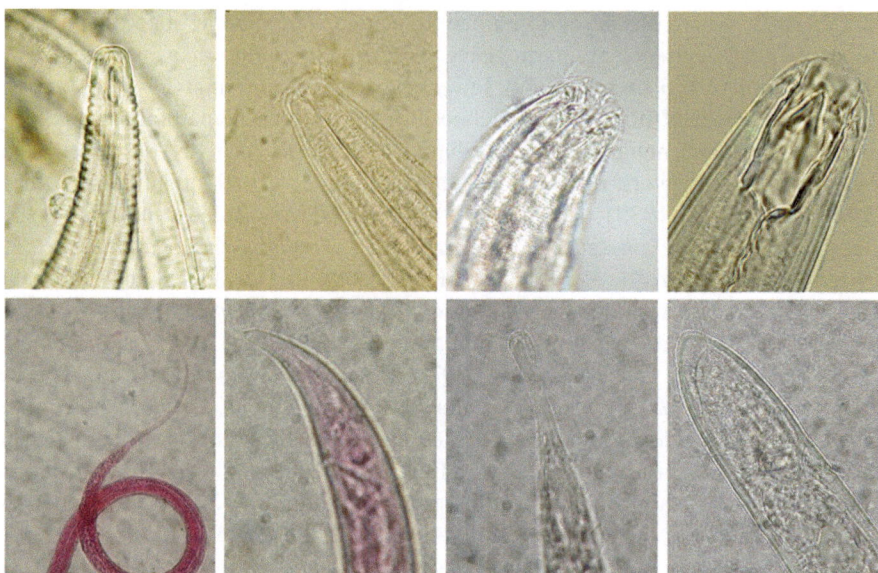

Fig. 3.1 Nematode morphological proxies used in functional trait determination; *top row*, *left* to *right*: feeding guild as inferred from buccal cavity structure (following classification of Wieser 1953); selective deposit feeders (Wieser's type **1A**); non-selective deposit feeders (**1B**); epistrate feeders (**2A**); predators-omnivores (**2B**); *bottom row*, *left* to *right*: tail shapes; filiform elongated; conical; clavate, conical-cylindrical; short rounded (*Photo* J. Rokicka-Praxmajer)

identified primarily from the morphology of their buccal cavity structure and armature (Jensen 1992; Wieser 1953). It is generally thought that most deep-sea nematodes belong to microphages (Wieser's groups 1a and 1b, Fig. 3.1), i.e. organisms feeding on microbes and organic matter these produce (Gage and Tyler 1991). The nematodes examined by Bussau et al. (1995) were dominated (66 % of all the species found) by microphages, followed (about 16 % of the species) by epistrate feeders the buccal cavity of which is adapted to scraping edible particles off hard surfaces. Fairly numerous (about 14 % of all the species) were also predators and scavengers. Particularly noteworthy in Bussau et al.'s (1995) material are nematodes, identified by them as representing the family Benthimermithidae, which lack buccal cavity (cf. Tchesunov 1997; Miljutin and Miljutina 2009b). The nematodes collected from the eastern part of CCFZ were also dominated by microphages (T. Radziejewska, personal observation; V.V. Galtsova, personal communication), and no nematodes lacking buccal cavity were found. However, as mentioned above, Miljutin and Miljutina (2009b) did find those nematodes in their material from CCFZ.

Regarding the distribution in the sediment, the deep-sea Pacific nematodes were shown to inhabit primarily the topmost 3 cm of sediment. Most studies revealed that layer to harbour up to 90 % of all the nematode individuals found in any core (Alongi 1992; Kaneko (Sato) et al. 1995; Shirayama 1984b; Snider et al. 1984, Radziejewska 2002), although single specimens were recorded as deep as 33 cm in the sediment (Shirayama 1984b). It is interesting to recall an observation by Bussau and Lorenzen (1991) who found nematodes in crevices of nodules retrieved from deep in the sediment (even as deep as 30–40 cm). In some studies, a sub-surface peak in nematode density, usually within the 0.5–1 or 1–2 cm layer, would appear (Alongi 1992; Shirayama and Kojima 1994; Snider et al. 1984; Radziejewska 2002). It is thought that, in the absence of human intervention, such a pattern of nematode (and meiofaunal in general) distribution is a result of natural disturbance inflicted on the seafloor surface by hydrodynamic processes (Aller 1997; Thistle et al. 1999) and /or by bioturbation caused by the activity of mobile epi- and megabenthos and discernible on the seafloor surface in the form of biogenic traces (the so-called *Lebensspuren*) (Gage and Tyler 1991).

Analyses of vertical distribution of nematodes in the sediment have revealed two other interesting effects. The first involves a trend towards increased body size with depth in the sediment (Gage and Tyler 1991; T. Radziejewska, personal observation); The trend, known also from the shelf habitats, e.g. off Louisiana in the Gulf of Mexico (Radziejewska, unpublished data) or the deep areas of the Mediterranean Sea (Soetaert and Heip 1989), is explained by reference to the fact that large specimens, being more mobile and physically stronger, are better equipped to penetrate deeper layers of fairly cohesive sediment in order to take advantage of energy reserves there. Ontogenetic migrations cannot be ruled out, either, but inferences are hampered by the absence of data on growth rates and developmental cycles of deep-sea nematodes. In the context of nematode size variability, it is interesting to invoke the finding of Vanreusel et al. (1997) that the nematodes inhabiting the sediment at the base of hydrothermal vents are, on the average, twice as large as those dwelling in fine abyssal sediments away from the vents. Vanreusel et al. (1997) tried to account for the difference by invoking faciliated transepidermal absorption of dissolved organic matter in the vicinity of hydrothermal vents, which, in turn, enhances the nematode growth rate.

The other aspect of variations in the vertical distribution of nematodes concerns taxon-specific sediment depth preferences within the meiobenthos. Those preferences are best known in shallow-water nematodes and harpacticoid copepods to the extent that workers have identified assemblages of nematode and harpacticoid species and genera typical of certain depths in sediment (Galtsova 1991; Joint et al. 1982). As pointed out by Joint et al. (1982), the taxon-specific preferences in question are much stronger in harpacticoids than in nematodes. So far, there has been no evidence supporting the downcore variability in nematode distribution in the abyssal sediments which could be related to taxon-specific depth preference. As the nematodes found in the uppermost 3 cm of the sediment collected at the IOM test site were identified to genus, an attempt was made—using the Average Living Depth (ALD) parameter of Jorissen et al. (1998) to find out if any such preferences could be detected within that layer and at the genus level of taxonomic discrimination. The analysis showed the available space to be partitioned such that some genera (e.g. *Desmoscolex*, *Pareudesmoscolex*) form densest concentrations in a relatively thin (0.5–1 cm thick) sediment layer; others (e.g. *Halalaimus*) concentrated within the sediment depth range of 0–1.5 cm; still others (e.g. *Aponema*) tended to concentrate somewhat deeper (1–2 cm), and there was a group (*Terschellingia, Longicyatholaimus, Theristus, Daptonema, Monhystera, Trochamus*) inhabiting a relatively thick sediment layer and reaching deeper (2–2.5 cm) in the sediment. In addition, the pattern of vertical distribution in the sediment does not seem to be stable in many genera. For example, in the eastern CCFZ study (Radziejewska 2002; Radziejewska et al. 2001), the sediment layer inhabited by some genera (*Paralongicyatholaimus, Daptonema, Comesoma, Acantholaimus, Diplopeltoides*) in 1997 was expanded compared to the thickness of the layer they occupied in 1995; a reverse—contraction of the layer of the densest occurrence—was true with respect to other genera (*Terschellingia, Longicyatholaimus, Linhystera,* and *Theristus*). It should be remembered that, due to the small number of samples and time points (see later) as well as the narrow sediment depth range analysed (0–3 cm), those findings have to be viewed with caution. They are not amenable to statistical testing, either.

On the other hand, it has been established that the abyssal sediment characteristics change along a vertical sediment profile, the major alteration of geotechnical properties being observed at the depth of about 3 cm in the sediment (Radziejewska and Modlitba 1999) at the bottom of the geochemically active layer. Most probably, the sediment (pore water) oxygen profile changes as well. It is then plausible that, similarly to the individual size and abundance of nematodes (see below), the taxonomic composition of an assemblage undergoes a change at that depth. To verify this contention, however, a detailed taxonomic study on the entire sediment depth taken into account in the meiofauna analysis would be necessary. Such a study would be highly desirable, because acquisition and partition of the available space by meiobenthic organisms is an interesting ecological question, one that is being currently in the focus of attention and discussed—for shallow-water areas—mainly from the standpoint of meiobenthic biodiversity drivers. The main emphasis is on the utilisation, by the meiofauna, of small-scale inhomogeneities in distribution of food resources (Moens et al. 1999), competitive interactions, predation, and natural disturbances in sediment structure (Varon and Thistle 1988) as well as species-specific tolerances of reduced oxygen levels (e.g. Steyaert et al. 2007). However, no attempt to tackle this problem has been made with respect to the abyssal meiobenthos.

The discussion on sediment properties-related taxonomic structure of the abyssal Pacific nematode taxocoene should not disregard another manifestation of habitat-dependent variability. The sedimentary environment in the deep sea is far from being homogenous, if only on account of the physical sediment structure (the presence of grains of various size, larger or smaller interstitial spaces). The inhomogeneity occurs at various spatial scales and is additionally affected by the presence of certain structures such as, e.g. polymetallic nodules. As shown by the few studies published so far (Bussau et al. 1995; Mullineaux 1987; Thiel et al. 1993), the nodules constitute an unusual biotope as they provide, in the semiliquid abyssal sediment, the only hard surface amenable to being colonised by sessile organisms from microbiota (Veillette et al. 2007a, b) to megafauna (e.g.

Kamenskaya et al. 2013). The nodule-inhabiting fauna in the Clarion-Clipperton Fracture Zone was described in detail by Mullineaux (1987, 1988) and Veillette et al. (2007a, b). They found the fauna to be mainly sessile, encrusting, qualitatively and quantitatively rich, and frequently forming a dense cover over a good part of a nodule's surface. They have provided evidence that the nodule fauna is a unique assemblage of species, in some respects similar to a shallow-water hard bottom community. The sessile organisms permanently attached to the nodule surface are primarily suspension feeders. Their feeding activity results in accumulation, on the nodule surface, of a microbiological film, i.e. a thin layer of organic matter and microorganisms. Some of those are capable of extracting metal ions from the near-bottom water layer and /or from the nodule surface (Verlaan 1992). The organic film and microbes (Burnett and Nealson 1981) as well as organisms attached to a nodule or living in numerous nooks and crannies on its surface contribute to trophic resources for mobile predators and scavengers among the macro- and megabenthos (Paul 1976).

According to Thiel et al. (1993), the crevices on the nodule surface, filled with sediment particles, are inhabited mainly by micro- and meiobenthos. Bussau et al. (1995) and Thiel et al. (1993) found substantial differences between the crevice-dwelling meiobenthic assemblages and those inhabiting the surrounding sediment. The material washed off the nodules surface and crevices was found to contain nematodes dominated by genera that were relatively rare in the surrounding sediment. The authors mentioned reported the nodule crevice nematodes to be dominated by *Acantholaimus* (family: Chromadoridae), *Paracyatholaimus* (Cyatholaimidae), *Molgolaimus* (Desmodoridae), *Camacolaimus* (Leptolaimidae), *Thalassomonhystera* (Monhysteridae), *Syringolaimus* (Ironidae), and *Trefusia* (Trefusiidae). Thiel et al. (1993) stressed that, in their collection, nematodes belonging to the families Cyatholaimidae, Desmodoridae, Leptolaimidae, Ironidae, and Trefusiidae were associated mainly with the nodules, while members of the Microlaimidae, Diplopeltidae, Desmoscolecidae, and Oxystominidae were primarily typical of the surrounding sediment.

In the study described by Radziejewska (2002) and Radziejewska et al. (2001), when sampling the meiobenthos in the eastern part of CCFZ, no separation between nodule and sediment fauna was made, mainly because the site was basically nodule-free. The few nodules found in multicorer tubes were picked out and washed, the washed-off sediment being placed together with the surface sediment layer (Radziejewska 2002). On the other hand, three multicorer deployment were made in a nodule-bearing area (Radziejewska and Modlitba 1999). For this reason, the discussion below pertaining to the nodule nematofauna is restricted to general differences between nematode assemblages found in the nodule-free and nodulised areas, the latter—by definition—containing also those nematodes inhabiting nodule crevices.

The analysis (Fig. 3.2) showed the three cores collected from the nodulised area to be, in terms of their nematode taxocoene composition, clearly different from the remaining cores. The nodule field nematodes were spread among 83 genera, 9 of which (*Trileptium, Ironella, Metaparoncholaimus, Oncholaimus,*

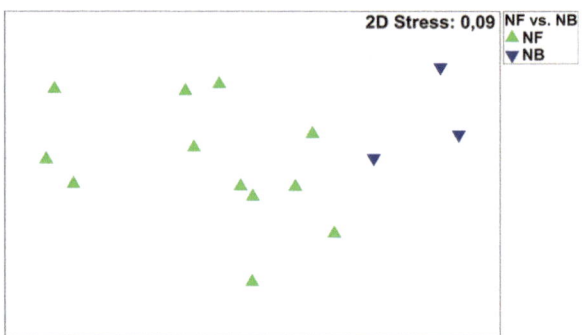

Fig. 3.2 A non-metric multi-dimensional similarity (NMDS) plot showing differentiation in taxonomic (genus-level) composition of nematode fauna in nodule-bearing (*NB*) and nodule-free areas (*NF*) of the non-disturbed bottom sampled in 1995 and 1997 in the studies described by Radziejewska (2002) and Radziejewska et al. (2001)

Steridora, Spirinia, Perepsilonema, Plectolaimus, Eleutherolaimus) were, in 1997, present only there, i.e. seemed to be associated with the presence of nodules in the sediment. However, only three of those genera (*Ironella, Metaparoncholaimus,* and *Perepsilonema*) could be regarded as nodule field-specific, as they were absent from the nodule-free sediment sampled in 1995. Intuitively, one would expect a higher diversity (genus richness) in a more complex habitat offered by the nodulised field. As shown, however, by the rarefaction curves and k-dominance plots published by Radziejewska et al. (2001), the nematode genus diversity seem to be lower in the nodulised area, compared to the nodule-free sediment: a lower number of dominant genera in the first was accompanied by a higher per cent domination in the taxocoene. This result, however, could stem from a far lower sampling effort in the nodulised area (3 cores), which precludes drawing any firm conclusions from these data alone. However, taxonomic evidence from other nodule-bearing area may help to elucidate the differences between nodule-free *versus* nodulised area. Such an analysis, for the central part of the CCFZ, was begun by Miljutina et al. (2010). As shown by their study, although both habitat types supported similar numbers of morphospecies (239 and 230 in the nodule-free and nodule-bearing sediments, respectively), the two habitat types shared 144 morphospecies (about 44 % of the total number found), indicating strong habitat-dependent differences. At the genus level, nematode communities of both habitat types were dominated by *Thalassomonhystera*, the sub-dominants being similar as well (*Acantholaimus* and *Theristus*), but their proportions differed between the two habitat types, *Thalassomonhystera* and *Theristus* contributing the most to the dissimilarity between the taxonomic composition of the two habitat types.

Copepoda Harpacticoida. As already mentioned, harpacticoid copepods usually rank second in abundance among the Pacific abyssal metazoan meiobenthos. In various studies, they accounted for from about 2–3 to about 13 % of the total meiofaunal abundance (Alongi 1992; Gage and Tyler 1991; Kaneko (Sato) et al.

1995, 1997; Renaud-Mornant and Gourbault 1990; Radziejewska 2002; Ahnert and Schriever 2001; Shirayama and Kojima 1994; Snider et al. 1984; Mahatma 2009). They are regarded as a taxon successful in the deep-sea in the sense that, as demonstrated by Thistle (2001), their depth-dependent reduction in abundance is less severe than that of other invertebrates, particularly the macrofauna.

As with other meiofaunal taxa, information on taxonomic composition of the deep-sea harpacticoid fauna is relatively scant (Seifried 2004). References to most of the data published prior to the 1970s were provided by Thistle (1978) who, when discussing distribution of harpacticoid copepods in the San Diego Trough, referred to studies carried out in the Peru-Chile (Becker 1972) and Aleutian Trenches (Jumars and Hessler 1976) and in the northern part of the Central Pacific (Hessler and Jumars 1974). There is also a number of purely taxonomic publications from that time, which contain descriptions of new species (e.g. Becker 1972; Becker and Schriever 1979). On the other hand, authors of most ecological papers did not published harpacticoid taxa lists. They did, however, emphasise—similarly to the case of nematodes—the high diversity (or taxonomic richness) of those copepods. For example, having processed sediment in 10 samples collected with a box corer at the depths of 5,500–5,800 m in the central North Pacific, Hessler and Jumars (1974) reported finding a total of 57 species; Jumars and Hessler (1976) identified 20 species in a single box core retrieved from the Aleutian Trench (7,298 m water depth) (Table 3.3). It should be pointed out that—similarly to nematodes—most harpacticoids found in the abyssal sediments belonged to new, undescribed species (putative species; Thistle 1978).

More recent data on composition of the abyssal Pacific harpacticoids can be found in papers resulting from German studies in the Peru Basin (Project DISCOL and subsequent programmes; Thiel 2001). Bussau et al. (1995) as well as Ahnert and Schriever (2001) found the harpacticoid community to be represented by members of 19 families, with the Ectinosomatidae, Ameiridae, Tisbidae, Argestidae, Neobradyidae, Diosaccidae, and Danielssenidae being dominants.

In the materials described by Radziejewska (2002) and Radziejewska et al. (2001), from the eastern part of CCFZ, harpacticoid copepods (identified in the top 3 cm of cores) contributed from about 2.5 to about 12 % to the total meiobenthos abundance. All samples showed a distinct presence of juvenile stages, the copepodites, which accounted for up to 45.5 % of all the 1,718 individuals subjected to taxonomic identification. Incidentally, as revealed by the most recent studies on deep-sea harpacticoids in the Atlantic, the predomination of copepodites seems to be a typical characteristic (G. Viets-Kohle, K.-H. George, P. Martinez Arbizu, personal communication; T. Radziejewska et al., unpublished conference paper). Mahatma (2009) reported that as few as 20 % of harpacticoids he examined were identifiable adult specimens. It has been only recently (Menzel 2011) that first descriptions of abyssal harpacticoid copepodites began to appear. In most of the eastern CCFZ copepodites, family characters were not developed strongly enough for the animals to be assigned to a family, let alone a genus. In addition, a considerable proportion of adults could not be identified to genus, either.

Table 3.3 Harpacticoid copepod diversity reported from the Pacific abyssal seafloor; BC, box corer; MC, multiple corer

Area	Depth range (m)	Source	Total no. of species/genera	Dominant families	Dominant species/ genera	Remarks
Central North Pacific	5,500–5,800	Hessler and Jumars (1974)	57 species	Not reported	Not reported	No mention of nodules on the seafloor
Peru Basin	4,037–4,158	Bussau et al. (1995), Ahnert and Schriever (2001)	Not reported	Cletodidae, Ectinosomatidae, Ameiridae, Tisbidae (Bussau et al. 1995)	Not reported	Nodule-bearing area
				Ectinosomatidae, Ameiridae, Tisbidae, Argestidae, Neobradyidae, Diosaccidae, Danielssenidae (Ahnert and Schriever 2001)		
CCFZ, eastern part	4,380–4,430	Radziejewska et al. (2001)	62 genera	Tisbidae, Ameiridae, Paramannopidae, Argestidae	*Tachidiopsis, Pseudomesochra, Zosime, Metahuntemannia, Eurycletodes, Sarsameira*, undescribed argestid	Both nodule-free (undisturbed and disturbed) and nodule-bearing areas
CCFZ, central part (French claim area; cf. Fig. 3.3)	4,947–5,046	Mahatma (2009)	61 genera	Ameiridae, Argestidae, Ectinosomatidae, Canthocamptidae	Not reported	Both nodule-bearing and nodule-free areas
CCFZ central part	4,860	Radziejewska et al. (unpublished)[a]	34 genera	Tisbidae, Argestidae, Miracidae (formely Diosaccidae)	*Mesocletodes, Tachidiopsis, Pseudotachidius, Cletodes*	Both nodule-bearing and nodule-free areas; harpacticoids identified in BC and MC samples mostly to 0.5 cm in the sediment, with additional data from 1.5 cm

[a]See Footnote to Table 3.1

In addition to the non-identifiable harpacticoids, those found within the surficial sediment in the eastern CCFZ consisted of representatives of 16 known families and 62 known genera (Table 3.3). Most identified individuals represented the families Tisbidae, Ameiridae, Paranannopidae, and Argestidae; the most abundantly represented genera included *Tachidiopsis, Pseudomesochra, Zosime, Metahuntemannia, Eurycletodes, Sarsameira,* and an undescribed argestid (Table 3.3).

In the study carried out in the central part of CCFZ (Trueblood et al. 1997), samples of the undisturbed sediment (the topmost 0.5–1.5 cm layer) revealed the presence of a total of 34 genera (Radziejewska et al. unpublished conference paper) (Table 3.3) representing primarily the families Tisbidae, Argestidae, and Miraciidae (reported as Diosaccidae, at present an invalid family name according to the World Register of Marine Species, www.marinespecies.org). In that study, copepodites contributed up to about 70 % of all harpacticoid specimens, so—like in other studies—the majority of the taxonocoene remained unidentified.

Various authors [e.g. Kaneko (Sato) et al. 1997; Radziejewska 2002; Snider et al. 1984] reported the harpacticoids to aggregate in the surficial sediment layer. As opposed to nematodes, they did not form subsurface peaks, possibly because, by analogy with shallow-water assemblages (Arlt et al. 1980; Palmer and Gust 1985), there are deep-sea harpacticoids which actively migrate from the sediment into the near-bottom water layer and back (Thistle et al. 2007a). These migrations are facilitated by morphological adaptions to swimming, involving the presence of numerous and long setae on mouth and thoracic appendages (Renaud-Mornant and Gourbault 1990; Thistle 1978). Such adaptations are particularly typical of the families Miraciidae (formerly Diosaccidae), Ectinosomatidae, and Thalestridae (Thistle and Sedlacek 2004), to mention those whose representatives are found on or close to the sediment surface in the deep sea. In contrast, there are also harpacticoids whose morphology indicates that they are capable of burrowing in the sediment so that they are usually found deeper downcore, in more compacted sediment layer. As opposed, however, to typically interstitial harpacticoids which are small-sized, with a worm-like elongated body and a low number of setae on the appendages (Giere 2009), the deeper-dwelling deep-sea harpacticoids have massive bodies and relatively large and strong appendages. Subsurface sediment dwellers are particularly common in the families Ancorabolidae, Cerviniidae, Canthocamptidae, and Huntemanniidae (I. Drzycimski, personal communication). Due to the fact that, in the material from the eastern part of CCFZ, taxonomic identification involved basically those harpacticoids found within the 0–3 cm layer, the data with which to compare different genera in terms of their vertical specialisation are not sufficiently abundant. There are, however, indications that such specialisation does exist, for which reason one may assume that various taxa will be vertically segregated, similarly to what has been observed in shallow water habitats (Joint et al. 1982). Ahnert and Schriever (2001) applied the concept of mean sediment depth (msd) (not to be confused with the Average Living Depth mentioned earlier in the analysis of nematode depth distribution) to explore the vertical distribution of harpacticoids (identified to the family level). They found most of their harpacticoids to concentrate within the top 2 cm of the sediment, but the msd values proved to be, to some extent, family-dependent: while the Argestidae, Paramesochridae, Tisbidae, and Miraciidae (=Diosaccidae) tended to live closer to the sediment surface (msd < 1.0 cm), the msd of Thalestridae, Paranannopidae, and Canthocamptidae exceeded 1.5 cm.

Adaptations to living at a particular depth in the sediment may be related to inter-specific differences in preferences, partitioning, and modes of utilisation of food resources encountered in the sediment and in the near-bottom water layer (Renaud-Mornant and Gourbault 1990; Thistle 1978), similarly to what is known from shallow water situations (e.g. Buffan-Dubau and Carman 2000; Decho and Fleeger 1988). Unfortunately, the data collected so far are too scant to pose testable hypotheses concerning those and other aspects of harpacticoid biology in the deep-sea.

Another group of factors influencing the composition, abundance, and distribution of harpacticoid taxocoene is related to physical habitat characteristics (e.g. bottom cover type, presence of biogenic structures, near-bottom flow regime). Although these have been investigated mostly in coastal areas, there is evidence that such factors affect the deep-sea harpacticoids as well. For example, exploring the role of various biogenic structures on the seafloor (mud balls formed by the cirratulid polychaetes *Tharyx luticastellus* and *Tharyx monilaris*, polychaete tubes, aglutinating foraminiferans of various shapes, tanaid crustacean tubes) in the distribution of harpacticoids, Thistle (1979, 1983) observed a significant association between the structures and the copepods, but no effect on harpacticoid diversity. Thistle (1998) followed effects of near-bottom flow on diversity of harpacticoid copepods and found the diversity to be higher at a site eroded daily by strong near-bottom flows than at a site experiencing episodic erosion. This suggests that the erosion regime may give rise to small-scale habitat heterogeneity that promotes harpacticoid diversity.

With respect to physical habitat effects on harpacticoid copepods, of a particular interest in the context of this work are effects of the presence or absence of nodules on the seafloor. This was explored in very few studies only. As in the nematodes, taxonomic composition of the harpacticoids found in the upper 3 cm of the sediment in the eastern part of CCFZ allowed to compare qualitative characteristics of the harpacticoid taxocoene between nodule-free and nodulised areas (Radziejewska et al. 2001). Although the genus-level similarity analysis (Fig. 3.3) showed some differentiation between assemblages inhabiting the two habitat types, the pattern, probably on account of a lower number of taxa entered into the analysis, is weaker than that shown by nematodes.

The data set in question showed the harpacticoid taxon richness to be higher in the undisturbed nodule-free habitat (altogether 51 identified genera, 12 cores) than in the nodule-bearing sediment outside of the test site (22 genera, 3 cores), but the nodulised area samples yielded 4 identified harpacticoid genera (*Pontostratoites, Stenocopia, Nannopus,* and *Microsetella*) absent from the nodule-free site. This discrepancy could, however, be a spurious effect associated with a low sampling effort with respect to nodule bearing area (three sites only) coupled with a generally lower abundance of harpacticoids than nematodes, in which the habitat structure-related effect was very distinct. In his study addressing comparisons

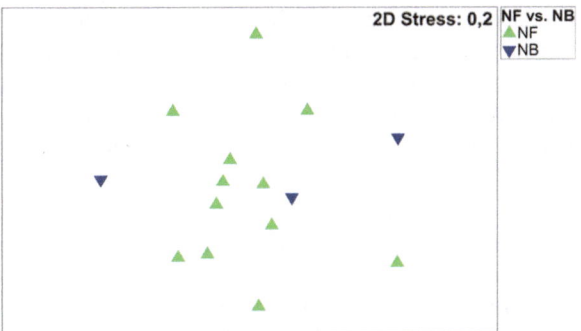

Fig. 3.3 A non-metric multi-dimensional similarity (NMDS) plot showing differentiation in taxonomic (genus-level) composition of harpacticoid copepods in nodule-bearing and nodule-free areas of the non-disturbed bottom sampled in 1995 and 1997 in the studies described by Radziejewska (2002) and Radziejewska et al. (2001)

between harpacticoid fauna in the CCFZ nodule-bearing and nodule-free areas with 5 stations each, Mahatma (2009) was able to observe a distinct separation, in terms of the taxonomic structure of harpacticoids, between the assemblages in the two habitat types. In addition to 26 shared genera, the nodule-free sites supported a total of 20 unique genera against 15 unique genera occurring on the nodule-bearing bottoms. Rarefaction curves presented by Mahatma (2009) show no asymptote, which suggests there would be still more genera living in both habitat types. Incidentally, the small set of nodulised bottom-specific harpacticoids in the easstern part of CCFZ (Radziejewska et al. 2001) shares only one genus, *Pontostratoites*, with the analogous part of the Mahatma's (2009) data set (assuming that the harpacticoid name was apparently misspelled in the relevant table he included).

As already pointed out, in addition to nematodes and harpacticoids, the abyssal meiobenthic communities in the Pacific include also other higher taxa, but—on account of their relatively poor representation in samples—none of them has received as much attention as the nematodes and harpacticoid copepods discussed above. The eastern CCFZ site samples contained such little-studied (at least in the Pacific deep-sea sediments) taxa as chaetonotid gastrotrichs, loriciferans, kinorhynchs or tardigrades. A few of such taxa were identified in the Pacific abyss [e.g. 3 species of tardigrades identified by Bussau et al. (1995) in the Peru Basin].

3.4 Biodiversity Issues

Some of diversity issues were already touched upon in the preceding paragraphs, but no discussion of faunal assemblages would be complete without a separate mention of diversity.

Biological diversity is a concept which encompasses both the number of species and/or of higher taxonomic units (species/taxon richness) and the partitioning of individuals between those units (evenness) (Hill 1973). It is an important attribute of communities and ecosystems, as it is a manifestation of natural complexity. Biological diversity is considered to be closely linked to the functioning of ecosystems (Cardinale et al. 2000) and has been documented to reflect phenomena affecting communities and ecosystems on various temporal and spatial scales (Angel 1996). It has to be mentioned that biodiversity, as a property of communities or ecosystems, can be expressed in different ways. The most common are two complementary approaches. One involves treating biological diversity as a metric (mathematical index) integrating information on both the number of taxa (e.g. species or genera) and the partitioning of individuals between those taxa. There has been a multitude of such indices proposed and used in ecological research (cf. Heip et al. 1988), some (e.g. Shannon-Wiener information and Margalef's indices, to mention those most commonly used) becoming routinely calculated to accompany species censuses. The other approach equals biodiversity with the number of species (taxa) alone and is termed the species (taxon) richness. The latter approach, because of its intuitive appeal, is extensively used and Boucher and Lambshead (1995) recommend the use of indices based on species richness for comparative purposes. As could be seen from the discussion in the preceding section, this latter approach (genus richness) is used here.

It is commonly maintained that the tropical rain forests and coral reefs are ecosystems characterised by the highest diversity in the world, their richness component being readily perceived even by a casual observer (Boucher and Lambshead 1995). The diversity of shelf faunas, too, has been relatively easy to assess. Observation of the diversity of life becomes more difficult as the observation field is extended over vast expanses of the deep sea. The sheer magnitude of those areas makes it tempting to produce sweeping generalisations, while such generalisations are open to criticism, precisely because of that reason (Gray 1994; Gray et al. 1997). Nevertheless, evidence accumulated for decades resulted in a widely held notion—raised to the status of a paradigm in the contemporary marine biology—that the deep-sea fauna is extremely diverse (Ramirez-Llodra et al. 2010). A particularly strong support to this notion has been provided by research on the deep-sea meiobenthos. The wealth of data Lambshead (1993 in Boucher and Lambshead 1995) obtained from the Porcupine Seabight (North-East Atlantic) allowed him to claim that the deep-sea floor sediment may support as many as 10^8 nematode species alone. In later publications (e.g. Boucher and Lambshead 1995; Lambshead et al. 2002, 2003), those estimates were brought to a lower, but still very high, level. The work done in the abyssal Pacific (see references in the preceding sections) fully supports the notion of the high diversity, at least at the local (alpha) level. This has been confirmed by theoretical considerations (Mokievsky and Azovsky 2002) and in the fundamental recent synthesis of deep-sea diversity published by Rex and Etter (2010), who, in their treatment of local diversity wrote that "At comparable densities, the deep-sea assemblages have about twice the number of species as do shallow water assemblages. It seems reasonable to conclude that for nematodes, the largest meiofaunal taxon, local deep-sea diversity is considerably higher than local coastal diversity" (p. 58).

Evidence presented in the preceding section attests to the high local diversity of both nematodes and harpacticoid copepods. In the latter, however, much of that diversity remains unresolved due to the overwhelming presence of juvenile specimens in the samples and an apparently strong area-taxonomic richness relationship (or, perhaps, genus turnover between the sites).

It has been debated whether the high local (alpha) diversity, the most readily observable diversity measure, can be translated into equally high beta (and higher level) diversity (taxon turnover between localities and areas). Rex and Etter (2010) emphasise the paucity of knowledge on beta diversity in the deep-sea meiofauna, which hampers drawing plausible conclusions regarding this aspect of diversity. With regard to the macrofauna, they by and large adhere to the "sink" hypothesis of Rex et al. (2005) stating that "much of the abyssal fauna is a more homogenous subset of the lower bathyal fauna" (p. 198). However, recent genetic analyses revealed a huge diversity of deep-sea macrobenthos even at regional scale (Brandt et al. 2007), thus contradicting the sink hypothesis. An even higher diversity is related to the microbial biosphere (Armstrong et al. 2012). It could then be suspected that the same could be true with respect to the abyssal meiobenthos. Although certain taxa, of both harpacticoids and nematodes, are cosmopolitan in their distribution in the abyssal realm (e.g. *Acantholaimus* in nematodes; Vanreusel et al. 2010) and, according to Seifried (2004), no truly deep-sea harpacticoid families could be singled out, many genera (e.g. *Cerviniopsis, Pontostratoites, Antarcticobradya, Marsteinia*) occur predominantly in the deep sea, and families such as the Argestidae, Zosimidae, Neobradyidae, or Aegisthidae are abundantly represented as well. Besides, and most importantly, virtually no species-level data exist, which is a great stumbling block for any more detailed analysis.

The debate, with respect to all biological compartments of the deep-sea benthic ecosystems, but particularly concerning the much lesser-known ones (the meio- and microbenthos), has practical aspects. Its possible outcomes would be of a great importance for planning a correct management of human interactions with deep-sea biodiversity, e.g. in a nodule field, guiding the design of a system of deep-sea areas exempt from development. Evidently, more such areas would be needed to conserve biodiversity should beta diversity prove high. In the absence of conclusive evidence, the precautionary principle seems appropriate, as evidenced in a recently established protected areas (termed the Areas of Particular Environmental Intereste or APEIs) in the CCFZ (Wedding et al. 2013).

3.5 Variability of Distribution of the Pacific Deep-Sea Meiobenthic Communities: Quantitative Characteristics

Data sets on meiobenthic abundance in the abyssal Pacific areas are much more numerous that are those on taxonomic composition. A comprehensive analysis of quantitative information on the deep-sea meiobenthos published before the 1970s was provided by Thiel (1983). The subsequent decades witnessed a marked increase in the amount of published data. However, the spatial coverage of this information is very uneven, and the knowledge base on the meiobenthos abundances and their changes in time and across the abyssal areas of importance for the future resource development (the CCFZ and Peru Basin) is not very broad.

The deep Pacific seafloor in areas known as nodule fields was found to support meiofaunal densities (Table 3.4) varying within a range on the order of 10^1–10^2 ind./10 cm^2, i.e. an order or two lower than the densities recorded in shallower and shelf waters (Giere 2009). Data in Table 3.4 seem to indicate a gear effect on abundance estimates: the estimates from sediment subsampled in the box corer content tend to be lower (sometimes very much so, cf. Ozturgut et al. 1978) than the estimates obtained from multicorer samples. This is again a strong indication that meiobenthos surveys in the deep sea should rely on sampling with a multicorer (Bett et al. 1994).

Most discussions on the abyssal meiobenthos abundance distribution point out to two effects underlying the differences observed: patchiness and the influence of sediment type.

In the deep-sea benthos, small scale patchiness was reported in a number of earlier studies (e.g. Bernstein and Meador 1979; Hessler and Jumars 1974). This type of patchiness is a typical feature of communities studied in coastal and shelf areas (Galtsova 1991; Giere 2009) and seems to be prevalent in the deep-sea meiobenthos as well. The small-scale patchiness is clearly evidenced by the wide range of densities recorded within a limited area (1.5 × 2 km) studied in the eastern part of CCFZ (Radziejewska 2002). Generally, explanations of meiobenthic patchiness in shallow areas have been sought by invoking non-uniform distribution of various environmental factors affecting meiobenthic animals. A particular attention has been paid to inhomogeneities in the physical environment (sediment and interstitial spaces) and the related chemical gradients (e.g. horizontal and vertical changes in oxygen content) as well as to the patchy distribution of food resources (Decho and Fleeger 1988; Galtsova 1991; Giere 2009). Owing to their small size, meiobenthic animals are able to perceive environmental gradients realised at very small spatial scales, and it is plausible to contend that this ability is maintained in the deep sea as well.

Physical and geochemical habitat inhomogeneities emerge from interactions and processes involving also the activity of other organisms—macro- and megabenthos. For example, the presence of xenophyophorids (Xenophyophorea), the giant protozoans along with their accompanying aggregates of organic matter, contributes markedly to inhomogeneities in the physical sediment structure (Levin 1991), increases the total sedimentary organic matter content (Levin and Gooday 1992), and creates peculiar "islands" or patches of discoloured sediment around the xenophyophorids. As a result, densities of both meio- and macrobenthos are observed to increase in the vicinity of those organisms (Levin 1991; Levin et al. 1986). Such local increases could have also been the case in the eastern CCFZ site where the meiobenthos was sampled, as the underwater images analysed by Radziejewska and Stoyanova (2000) showed a relatively high abundance of the xenophyophorids and their non-homogenous distribution. In addition,

Table 3.4 Abundance estimates of meiobenthos in the Pacific areas of importance for nodule deposit development, as produced by various studies

Area	Depth (m)	Source	Gear type (BC, box corer; MC, multicorer)	Total meiobenthos abundance (inds./10 cm²)	Remarks
CCFZ, central and eastern part	4,300–5,100	Ozturgut et al. (1978)	BC	10.8–32.2	baseline survey; averages recalculated from data reported
NE Pacific	5,821–5,874	Snider et al. (1984)	BC	91	Average
CCFZ, western part	4,960–5,140	Renaud-Mornant and Gourbault (1990)	BC	23–189	
Peru Basin	4,100–4,200	Bussau et al. (1995)	MC	76–216	Non-disturbed bottom
CCFZ, western part	5,315–5,330	Kaneko(Sato) et al. (1997)	MC	39–132	Non-disturbed bottom
CCFZ, central part		Trueblood et al. (1997), Trueblood et al. (1997)	BC, MC	46.2–205.3 (1993) 40.74–91.28 (1994)	BC used in 1993; MC used in 1994
CCFZ, eastern part	4,349–4,480	Radziejewska (2002)	MC	61.18–394.24	Non-disturbed, nodule-free bottom
CCFZ, eastern part	4,409–4,440	Radziejewska et al. (2001)	MC	137.97 (±3.49)–203.89 (±66.58)	Mean abundances; non-disturbed bottom sampled in 2000
CCFZ, eastern part	4,380–4,410	Radziejewska and Modlitba (1999)	MC	91.56 ± 23.31	Nodule-bearing area
CCFZ, eastern and central parts	4,830–5,091	Kim et al. (2004)	MC	53–113	
CCFZ, central part	4,966–5,027	Mahatma (2009)	MC	56.51–124.15	

inhomogeneity of the sedimentary milieu may also result from the presence of other large protists, the Komokiacea (Kamenskaya et al. 2012, 2013) which greatly affect the sediment texture. The Komokiacea and their fragments were found in abundance in the eastern part of CCFZ (Kamenskaya et al. 2012).

Sediment structure inhomogeneity results also from the activity of larger motile and sessile benthic organisms (macro- and megabenthos) and biogenic structures (e.g. tubes, burrows, pits) they produce in the sediment, i.e. the so-called *Lebensspuren* (Gage and Tyler 1991; Radziejewska and Stoyanova 2000). Such structures were observed in abundance in the area where meiobenthic samples were collected in the eastern part of CCFZ (Radziejewska and Stoyanova 2000; Tkatchenko and Radziejewska 1998). Thus, when interpreting meiobenthic patchiness on the deep seafloor, one has to refer to sedimentary inhomogeneities of the kinds mentioned above and treat them as manifestations of natural variability in the sedimentary environment. Such variability results in changes in the sediment structure, in fluxes of material from the near-bottom water and the sediment, and in the transport of solutes and solutions to and from the sediment. It can be assumed the small-scale meiobenthic patchiness reflects also those variations.

Variability of the meiobenthic communities in the shallow water is observed also on the temporal axis, as the meiobenthic densities have been frequently found to fluctuate in time (e.g. Coull 1985). The corresponding data from the abyssal depths are very limited, and most were provided by samples collected during environmental assessment programmes which will be discussed in Chap. 4. Information regarding seasonal—or longer-term (>year) variability in abyssal meiobenthos abundance and composition is near-absent (but see Kalogeropoulou et al. 2010). It may be presumed, however, that, similarly to other benthic organism categories (macro- and megabenthos), seasonal and temporal changes are an inherent characteristic of abyssal meiobenthic communities as well. It may be also contended that natural variability in the abyssal meiobenthic communities is induced by episodes of high hydrodynamic forcing on the bottom ("benthic storms") and those of phytodetritus sedimentation and other food falls (e.g. Smith and Baco 2003). With respect to scouring effect of strong near-bottom current, some light on the problem was shed by observations made in the easter CCFZ site site in 1997 (cf. Chap. 4). The sharp grooves left in the sediment by the device used to artificially disturb the sediment 22 months earlier were seen to have been smoothed out and eroded, which was most probably an effect of intensified bedload transport induced by the current (Tkatchenko and Radziejewska 1998). Such intensified hydrodynamics must have impacted on meiobenthic communities as well, as observed elsewhere in the deep sea, e.g. at the benthic-storm affected HEBBLE site off the Nova Scotia Rise (4,428 m depth) in the Atlantic (Thistle et al. 1991) where the meiobenthos, particlarly harpacticoids, were considerably affected by the intensified near-bottom current. The actual data from the Pacific are, however, non-existent.

Particularly interesting and important for the evolution of meiobenthic communities are effects of the deposition and presence of phytodetritus in the surficial sediment layers. Sedimentation of phytal material produced in the euphotic part of the water column down to the abyssal depths was first reported from the northern and equatorial Atlantic (Billett et al. 1983; Lampitt 1985; Thiel et al. 1988/1989). The phenomenon was subsequently intensely studied, primarily on the deep Atlantic floor (Gooday 1996; Pfannkuche et al. 1999; Rice et al. 1994; Soltwedel 1997), but also in the deep-sea Antarctic regions (Fileman et al. 1998) and in the Pacific (Beaulieu 2002; Beaulieu and Smith 1998; Lauerman et al. 1997; Loubere 1998; Smith et al. 1996, 1997). Periodic deposition of phytodetritus occurring in the form of aggregates of plant material consisting of both degraded and intact (fresh) phytoplankton cells takes place primarily and most intensively in temperate latitudes after the spring phytoplankton bloom. Nevertheless, phytodetritus sedimentation and deposition on the abyssal seafloor happens in the equatorial and sub-equatorial, oligotrophic areas of the oceans where seasonality of the upper part of the water column is much less pronounced (Beaulieu 2002).

Biochemical sediment assays performed after a phytodetritus sedimentation episode demonstrated the surficial sediment layer to become enriched in chloroplastic pigments, both the products of chlorophyll *a* degradation and chlorophyll *a* itself, thus providing evidence of a fairly fast sedimentation of the plant material down to abyssal depths. It was calculated that the plant material would reach the abyssal seafloor 3–6 weeks after the surface bloom (Drazen et al. 1998; Khripounoff et al. 1998). Smith et al. (1996) reported that chloroplastic pigment equivalents (CPE) in the central Pacific sediments at depth of 4,250–4,990 m ranged within 0.36–1.6 μg/g dry sediment, the chlorophyll *a* contribution ranging from 0 to 4.9 %. At one of the stations, they found a macrofaunal burrow at the depth of 24 cm in the sediment containing an accumulation of plant material in which CPE amounted to as much as 98 μg/g wet sediment weight! In comparison, the CPE of surficial sediment at abyssal depths of moderate latitudes of North Atlantic did not exceed 2.4 μg/g (Rice et al. 1986).

Sediment enrichment with allochthonous phytodetritus exerts a strong influence on the chemical composition of pore water, including the vertical profiles of dissolved oxygen content and contents of ions such as NO_3^- and Mn^{2+}; the exchange of solutes and solutions between the sediment and the near-bottom water is affected as well (Gehlen et al. 1997).

The phytodetritus deposited onto the seafloor surface and buried deeper into the sediment increases the sedimentary organic matter pool, thus augmenting food resources for abyssal heterotophs: microorganisms (Loubere 1998) and metazoan detritivores among both meio- (Soltwedel 1997), macro- (Smith et al. 1996, 1997), and megabenthos (Lauerman et al. 1997; Smith et al. 1997). However, while the phytodetrital enrichment was followed by distinct changes in the macro- and megafauna communities involving a dramatic increase in the abundance of one or more species (cf. the "*Amperima* event" in Billett et al. 2010) and might be inferred to provide a definite trigger for such a change (Radziejewska and Stoyanova 2000), evidence among the meiobenthic communities is less clearcut. For example, Lambshead and Gooday (1990) and Pfannkuche (1993) reported a lack of any distinct impact on metazoan meiofauna in the deep Atlantic, but recorded a rapid response among the deep-sea foraminiferans. In contrast, Brown

Fig. 3.4 An intact diatom cell (indicated by an *arrow*) among sediment particles at a station enriched by phytodetritus deposition in the eastern part of CCFZ sampled in 1997 (*Photo* T. Radziejewska)

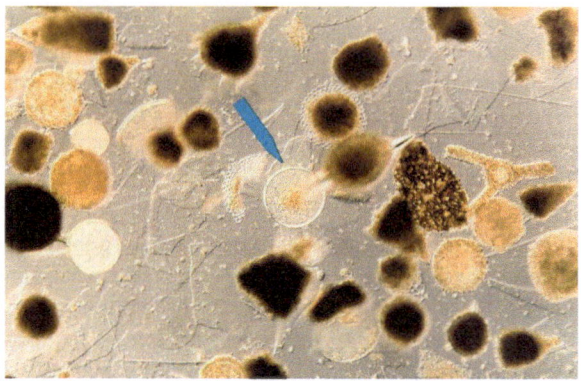

et al. (2001) observed the nematode abundance and biomass in the surface sediment layer at phytodetritus-enriched stations in the abyssal central North Pacific to be significantly higher than at non-enriched sites. In his review of biological responses to phytodetritus deposition on the deep seafloor, Gooday (2001) noted the metazoan meiofauna to be generally less responsive than protists (foraminiferans). He did, however, admit instances of positive responses among the metazoan meiobenthos. Such a response was observed in the eastern part of the CCFZ (Radziejewska 2002).

Some of the sediment cores collected in 1997 from the IOM BIE test site showed the presence of a very fluffy, light-weight fine material in the form of amorphous greenish-brownish aggregations (Radziejewska 2002 and personal observation). That material was very unevenly distributed on the sediment surface, as it was present at some of the sampling stations only and only in some of the cores of a respective multicorer array. At one of the stations, that material was particularly abundant and present in almost all of the cores. As already mentioned in Chap. 2, the fluorometric assay of the sediment from two cores revealed CPE values of 3.07 and 4.75 µg/g sediment dry weight (Radziejewska 2002). In one of the samples, 2.7 % of the total pigment content were accounted for by chlorophyll *a*, indicating an abundant and freshly deposited phytoplankton biomass. When viewed under the microscope, the sediment revealed a particular abundance of plant material, including intact diatom cells with green chloroplasts (T. Radziejewska, personal observation; Fig. 3.4).

The phytodetritus deposited on the sediment surface is patchily distributed and accumulates in microrelief depressions of the sediment surface (T. Radziejewska, personal observation). Therefore, effects of this material should be expected to be very patchy as well, producing inhomogeneities in distributions of biogeochemical microgradients and, in consequence, contributing to patchy distribution of fauna in the sediment.

The meiobenthos density (about 364 ind./10 cm^2) at the station with the sediment particularly heavily enriched with the phytal material turned out to be the highest in the entire series of the eastern CCFZ samples, thus providing indirect evidence of phytodetritus input effects on the meiofauna. In addition, the composition of the meiofauna at that station revealed some interesting details: the nematodes were strongly dominated (71.6 % of all the nematodes in the core!) by very small representatives of the family Desmoscolecidae (primarily from the genera *Desmoscolex* and *Pareudesmoscolex*; Radziejewska et al. 2001). For comparison, in 1995, representatives of the family contributed as little as 6.36 % of all the nematodes, while their overall contribution in 1997 increased to 20.20 % (Radziejewska 2002). The desmoscolecid nematodes were concentrated in the uppermost sediment layer. As already mentioned, Vanaverbeke et al. (2004), in their study on the shallow water meiobenthos off the North Sea coast, reported enhanced densities of small desmoscolecids as a result of a phytodetritus enrichment episode in the sediment.

The effects of freshly sedimented phytodetritus in the eastern CCFZ site sediment in 1997 were visible also among the harpacticoid copepods. Their density at the most heavily enriched station was particularly high as well, and the assemblage was dominated by adult females of a non-described (new) genus of the family Argestidae (I. Drzycimski, personal communication; Radziejewska et al. 2001), concentrated in the topmost sediment layer.

Most probably, the influx of sedimenting phytodetritus to the abyssal depths plays a role that can be likened to the importance of organic enrichment for the shallow water benthos. Benthic communities in organically enriched sediments of shallow water areas are usually strongly dominated by one or a few species the members of which are very small and reproduce rapidly, and are termed *r*-strategists or opportunist species (Pearson and Rosenberg 1978) which are also the earliest colonizers of sediments recovering from a drastic organic pollution (Pearson and Rosenberg 1978). Those species are believed to respond very rapidly to the enhancement of food resources. In the deep-sea case described, both desmoscolecid nematodes and argestid harpacticoids could be regarded as analogues of the opportunist species of the Pearson and Rosenberg (1978) model.

The observations described above suffer of a serious drawback in that they are incidental at best, for which reason effects of sediment changes on meiobenthic community structure are not amenable to any formal statistical testing. Nevertheless, those observations, although in need of additional evidence, could be useful in unraveling causes of temporal fluctuations and spatial variations in distribution of the abyssal meiobenthic communities and temporal variations they may undergo.

Spatial changes in quantitative characteristics of meiobenthic communities may be, similarly to qualitative features, also a result of differences in the properties of sedimentary cover. The major difference in CCFZ is whether or not the sediment is covered by polymetallic nodules. Radziejewska and Modlitba (1999) compared meiobenthic communities in the two habitat types and found marked differences paralleling differences in the lithological and geotechnical properties of the sediment. Although in both cases the sediment can be described as siliceous ooze composed (up to 75 %) of radiolarian skeletons, diatom frustules, and hexactinellid sponge needles, the nodule field sediment—in addition to being covered by polymetallic nodules—contained a higher proportion of coarser fractions

(>0.063 mm), due mainly to an increased contribution of micronodules. In terms of geotechnical properties, the differences between the two types of areas occurred primarily in the surface sediment layer, down to the depth of about 3 cm: the nodule field sediment was less water-logged, was characterised by lower plasticity, and showed also a much higher shear strength. The basic differences between the two habitat types were also evident in the subsurface layer and increased with depth in the sediment. Generally, the nodule field sediment proved coarser, more cohesive, and less plastic than that collected from the nodule-free area. As demonstrated by Radziejewska and Modlitba (1999), the nodule field meiobenthos was significantly less abundant (the mean density of 91.56 ± 23.31 ind./10 cm^2; Table 3.4) than that in the nodule-free sediment, the lower abundances being found in all the layers downcore. A similar finding—a significantly lower mean abundance of nematodes in the nodule-bearing than in the nodule free bottom (69 ± 19 and 137 ± 28 ind./10 cm^2, respectively)—was reported by Miljutina et al. (2010).

The information presented in this chapter forms a basis with which to view results of research involving the meiobenthos in the assessment of future anthropogenic impacts on the Pacific abyssal seafloor.

References

Ahnert A, Schriever G (2001) Response of abyssal Copepoda Harpacticoida (Crustacea) and other meiobenthos to an artificial disturbance and its bearing on future mining for polymetallic nodules. Deep Sea Res II 48:3779–3794

Aller JY (1997) Benthic community response to temporal and spatial gradients in physical disturbance within a deep-sea western boundary region. Deep-Sea Res I 44:39–69

Alongi DM (1992) Bathymetric patterns in deep-sea benthic communities from bathyal to abyssal depths in the western South Pacific (Solomon and Coral Seas). Deep-Sea Res 39:549–565

Angel M (1996) Oceanic biodiversity: origins and maintenance. In: Albertelli G, DeMaio A, Picasso M (eds) Atti 11° Congr Assoc Ital Oceanol Limnol. A.I.O.L., Genova, pp 33–60

Angel MV, Rice TL (1996) The ecology of the deep ocean and its relevance to global waste management. J Appl Ecol 33:915–926

Arlt G, Radziejewska T, Rodbertus L (1980) Vorlaufige Mitteilung über Untersuchungen zur Vertikalwanderung der Meiofauna in einem Flachwassergebiet der Darß-Zingster Boddenkette. Wissenschaft Zeitschr Wilhelm-Pieck-Univ Rostock 29(4/5):123–125

Armstrong C, Foley N, Tinch R et al (2012) Services from the deep: steps towards valuation of deep sea goods and services. Ecol Serv 2:2–13

Barnett PRO, Watson J, Connelly O (1984) A multiple corer for taking virtually undisturbed samples from shelf, bathyal and abyssal sediments. Oceanol Acta 7:399–408

Beaulieu SE (2002) Accumulation and fate of phytodetritus on the sea floor. Oceanog Mar Biol Ann Rev 40:171–232

Beaulieu S, Smith KL (1998) Phytodetritus entering the benthic boundary layer and aggregated on the sea floor in the abyssal NE Pacific: macro- and microscopic composition. Deep-Sea Res II 45:781–815

Becker K-H (1972) Eidonomie und Taxonomie abyssaler Harpacticoidea (Crustacea, Copepoda). PhD thesis, Christian-Albrechts-University, Kiel

Becker KH, Schriever G (1979) Eidonomie und Taxonomie abyssaler Harpacticoida (Crustacea, Copepoda) Teil III. 13 Neue Tiefsee-Copepoda Harpacticoidea der Familien Canuellidae,

Cerviniidae, Tisbidae, Thalestridae, Diosaccidae und Ameiridae. "Meteor"-Forsch Ergebn D 31:38–62

Belayev TM (1989) Glubokovodnye Okeanicheskye Zheloba I Ikh Fauna. Nauka, Moskva

Bernstein BB, Meador JP (1979) Temporal persistence of biological patch structure in an abyssal benthic community. Mar Biol 51:179–183

Bett BJ, Vanreusel A, Vincx M et al (1994) Sampler bias in the quantitative study of deep-sea meiobenthos. Mar Ecol Prog Ser 104:197–203

Billett DSM, Bett BJ, Reid WDK et al (2010) Long-term change in the abyssal NE Atlantic: The '*Amperima* Event' revisited. Deep Sea Res II 57:1406–1417

Billett DSM, Lampitt RS, Rice AL et al (1983) Seasonal sedimentation of phytoplankton to the deep sea benthos. Nature 302:520–522

Bluhm H (1994) Monitoring megabenthic communities in abyssal manganese nodule sites of the east Pacific Ocean in association with commercial deep-sea mining. Aquat Conserv Mar Freshw Ecosyst 4:187–201

Boucher G, Lambshead PJD (1995) Ecological biodiversity of marine nematodes in samples from temperate, tropical and deep-sea regions. Conserv Biol 9:1594–1604

Brandt A, Gooday AJ, Brandão SN et al (2007) First insights into the biodiversity and biogeography of the Southern Ocean deep sea. Nature 447:307–311

Brown CJ, Lambshead PJD, Smith CR et al (2001) Phytodetritus and the abundance and biomass of abyssal nematodes in the central, equatorial Pacific. Deep-Sea Res I 48:555–565

Buffan-Dubau E, Carman KR (2000) Diel feeding behavior of meiofauna and their relationships with microalgal resources. Limnol Oceanog 45:381–395

Burnett BR, Nealson KH (1981) Organic films and microorganisms associated with manganese nodules. Deep-Sea Res 28A:637–645

Bussau C (1993) Taxonomische und ökologische Untersuchungen an Nematoden des Peru-Beckens. PhD thesis, Christian-Albrechts-University, Kiel

Bussau C, Lorenzen P (1991) Die Nematodenfauna der ozeanischen Tiefsee—erste taxonomische Bestandsaufnahme. Proc Germ Zool Soc Stuttgart, 441–442

Bussau C, Vopel K (1999) New nematode species and genera (Chromadorida, Microlaimidae) from the deep sea of the eastern tropical South Pacific (Peru Basin). Ann Naturhist Mus Wien 101B:405–421

Bussau C, Schriever G, Thiel H (1995) Evaluation of abyssal metazoan meiofauna from a manganese nodule area of the eastern South Pacific. Vie Milieu 45:39–48

Cardinale BJ, Nelson K, Palmer MA (2000) Linking species diversity to the functioning of ecosystems: on the importance of environmental context. Oikos 91:175–183

Copley J, Cuvelier D, Desbruyéres D et al (2010) Temporal change in deep-sea benthic ecosystems: a review of the evidence from recent time-series studies. Adv Mar Biol 58:1–95

Costello MJ, Coll M, Danovaro R et al (2010) A census of marine biodiversity knowledge, resources, and future challenges. PLoS ONE 5(8):e12110

Coull BC (1985) Long-term variability of estuarine meiobenthos: an 11 year study. Mar Ecol Prog Ser 24:205–218

Decho AW, Fleeger JW (1988) Microscale dispersion of meiobenthic copepods in response to food-resource patchiness. J Exp Mar Biol Ecol 118:229–243

Decraemer W, Gourbault N (1997) Deep-sea nematodes (Nemata, Prochaetosomatinae): new taxa from hydrothermal vents and a polymetallic nodule formation of the Pacific (East Rise: North Fiji and Lau Basins; Clarion-Clipperton fracture zone). Zool Scripta 26:1–12

Dinet A, Desbruyères D, Khripounoff A (1985) Abondance des peuplements macro- et méiobenthiques: répartition et stratégie d'échantillonage. In: Laubier L, Monniot C (eds) Peuplements Profonds du Golfe de Gascogne. Inst Français Rech Exploit Mer, Paris, pp 121–142

Dinet A, Grassle F, Tunnicliffe V (1988) Prèmieres observations sur la méiofaune des sites hydrothermaux de la dorsale Est-Pacifique (Guaymas, 21°N) et de l'Explorer Ridge. Oceanol Acta Vol Spec 8:7–14

Drazen JC, Baldwin RJ, Smith KL Jr (1998) Sediment community response to a temporarily varying food supply at an abyssal station in the NE Pacific. Deep-Sea Res II 45:893–913

Etter RJ, Grassle JF (1992) Patterns of species diversity in the deep-sea as a function of sediment particle size diversity. Nature 360:576–578

Fasham MJR, Baltiño BM, Bowles MC (eds) (2001) A new vision of ocean biogeochemistry after a decade of the Joint Global Ocean Flux Study (JGOFS). Ambio Spec Rep 10:4–30

Fileman TW, Pond DW, Barlow RG et al (1998) Vertical profiles of pigments, fatty acids and amino acids: evidence for undegraded diatomaceous material sedimenting to the deep ocean in the Bellingshausen Sea, Antarctica. Deep-Sea Res I 45:333–346

Flach E, Vanaverbeke J, Heip C (1999) The meiofauna: macrofauna ratio across the continental slope of the Goban Spur (northeastern Atlantic). J Mar Biol Ass UK 79:233–241

Gage JD, Tyler PA (1991) Deep-sea biology: a natural history of organisms at the deep-sea floor. Cambridge University Press, Cambridge

Galéron J, Sibuet M, Vanreusel A et al (2001) Temporal patterns among meiofauna and macrofauna taxa related to changes in sediment geochemistry at an abyssal NE Atlantic site. Prog Oceanog 50:303–324

Galtsova VV (1991) Meiobenthos in marine ecosystems (with special reference to freeliving Nematodes). Proc Zool Inst USSR Acad Sci, St. Petersburg

Galtsova VV, Kamenskaya OE (1993) K metodam issledovaniya glubokovodnogo meiobentosa. In: Kuznetsova AP, Sokolova MM (eds) Pitaniye Morskikh Bezpozvonochnykh v Raznykh Vertikalnykh i Shirotnykh Zonakh. Inst Okeanol P.P. Shirshov Ross Akad Nauk, Moskva

Gehlen M, Rabouille C, Ezat U et al (1997) Drastic changes in deep-sea sediment porewater composition induced by episodic input of organic matter. Limnol Oceanog 42:980–986

Giere O (2009) Meiobenthology: the microscopic fauna in aquatic sediments, 2nd edn. Springer, Berlin

Glover AG, Gooday AJ, Bailey DM et al (2010) Temporal change in deep-sea benthic ecosystems: a review of the evidence from recent time-series studies. Adv Mar Biol 58:1−95

Gooday AJ (1996) Epifaunal and shallow infaunal foraminiferal communities at three abyssal NE Atlantic sites subject to differing phytodetritus input regimes. Deep Sea Res I 43:1395–1421

Gooday AJ (2001) Biological responses to seasonally varying fluxes of organic matter to the ocean floor: a review. J Oceanog 58:305–332

Gooday AJ (2002) Organic-walled allogromiids: aspects of their occurrence, diversity and ecology in marine habitats. J Foram Res 32:384–399

Gooday AJ, Pfannkuche O, Lambshead PJD (1996) An apparent lack of response by metazoan meiofauna to phytodetritus deposition in the bathyal north-eastern Atlantic. J Mar Biol Ass UK 76:297–310

Gooday AJ, Bowser SS, Cedhagen T et al (2005) Monothalamous foraminiferans and gromiids (Protista) from western Svalbard: a preliminary survey. Mar Biol Res 1:290–312

Gray JS (1994) Is deep-sea species diversity really so high? Species diversity of the Norwegian continental shelf. Mar Ecol Prog Ser 112:205–209

Gray JS, Poore GCB, Ugland KI et al (1997) Coastal and deep-sea benthic diversities compared. Mar Ecol Prog Ser 159:97–103

Hannides AK, Smith CR (2003) The northeastern Pacific abyssal plain. In: Black KD, Shimmield GB (eds) Biogeochemistry of marine systems. Blackwell, Oxford

Hebert PD, Gregory TR (2005) The promise of DNA barcoding for taxonomy. System Biol 54:852–859

Heip C, Herman PMJ, Soetaert K (1988) Data processing, evaluation and analysis. In: Higgins RP, Thiel H (eds) Introduction to the study of meiofauna. Smiths Inst Press, Washington, D.C.

Hessler RR, Jumars PA (1974) Abyssal community analysis from replicate box cores in the central North Pacific. Deep-Sea Res 21:185–209

Hessler RM, Kaharl VA (1995) The deep-sea hydrothermal vent community: an overview. In: Humphries S, Mullineaux L (eds), Seafloor hydrothermal systems: physical, chemical, biological and geological interactions. Am Geophys Union Geophys Monogr 91:72–84

Hill MO (1973) Diversity and evenness: a unifying notation and its consequences. Ecology 54:427–432

ISA (2007) Biodiversity, species ranges, and gene flow in the Abyssal Pacific Nodule Province: predicting and managing the impacts of deep seabed mining. ISA Techn Stud 3, International Seabed Authority, Kingston, Jamaica

Jensen P (1992) Predatory nematodes from the deep-sea: description of species from the Norwegian Sea, diversity of feeding types and geographical distribution. Cah Biol Mar 33:1–23

Joint IR, Gee JM, Warwick RM (1982) Determination of fine-scale vertical distribution of microbes and meiofauna in an intertidal sediment. Mar Biol 72:157–164

Jorissen FJ, Wittling I, Peypouquet JP et al (1998) Live benthic foraminiferal faunas off Cap Blanc, NW Africa: community structure and microhabitats. Deep-Sea Res I 45:2157–2188

Jumars PA, Hessler RR (1976) Hadal community structure: implications from the Aleutian Trench. J Mar Res 34:547–560

Kalogeropoulou V, Bett BJ, Gooday AJ et al (2010) Temporal changes (1989–1999) in deep-sea metazoan meiofaunal assemblages on the Porcupine Abyssal Plain, NE Atlantic. Deep Sea Res II 57:1383–1395

Kamenskaya OE, Gooday AJ, Radziejewska T et al (2012) Large, enigmatic foraminiferan-like protists in the eastern part of the Clarion-Clipperton Fracture Zone (abyssal north-eastern subequatorial Pacific): biodiversity and vertical distribution in the sediment. Mar Biodiv 42:311–327

Kamenskaya OE, Melnik VF, Gooday AJ (2013) Giant protists (xenophyophores and komokiaceans) from the Clarion-Clipperton ferromanganese nodule field (eastern Pacific). Biol Bull Rev 3:388–398

Kaneko (Sato) T, Ogura K, Fukushima T (1995) Preliminary results of meiofauna and bacteria abundance in an environmental impact experiment. In: Yamazaki T, Aso K, Okano Y et al (eds) Proceedings of ISOPE-Ocean mining symposium, Tsukuba, Japan, pp 181–186

Kaneko (Sato) T, Maejima Y, Teishima Y (1997) The abundance and vertical distribution of abyssal benthic fauna in the Japan Deep-Sea Impact Experiment. In: Chung JS, Das BM, Matsui T et al (eds), Proceedings of 7th ISOPE Conference, vol 1. Honolulu, USA, pp 475–480

Khripounoff A, Vangriesheim A, Crassous P (1998) Vertical and temporal variations of particle fluxes in the deep tropical Atlantic. Deep-Sea Res I 45:293–316

Kim D-S, Min W-G, Lee K-Y et al (2004) Meiobenthic communities in the deep-sea sediment of the clarion-clipperton fracture zone in the Northeast Pacific. Ocean Polar Res 26:265–272

Kim HJ, Kim D, Yoo CM et al (2011) Influence of ENSO variability on sinking particle fluxes in the northeastern equatorial Pacific. Deep-Sea Res I 58:865–874

Kim HJ, Hyeong K, Yoo CM et al (2012) Impact of strong El Niño events (1997/1998 and 2009/2010) on sinking particle fluxes in the 10 °N thermocline ridge area of the northeastern equatorial Pacific. Deep-Sea Res I 67:111–120

Kontar EA, Sokov AV (1994) A benthic storm in northeastern tropical Pacific over the fields of manganese nodules. Deep-Sea Res 41:1069–1089

Krishnamurthy KP, Francis R (2012) A critical review on the utility of DNA barcoding in biodiversity conservation. Biodiv Conserv 8:1901–1919

Kristensen RM (1983) Loricifera, a new phylum with aschelminthes characters from the meiobenthos. Zeitschr Zool System Evolutionsforsch 21:163–180

Lambshead PJD (2011) Introduction. In: ISA, Marine Benthic Nematode Molecular Protocol Handbook (Nematode Barcoding). ISA Techn Stud 7, International Seabed Authority, Kingston, Jamaica

Lambshead PJD, Gooday AJ (1990) The impact of seasonally deposited phytodetritus on epifaunal and shallow infaunal benthic foraminiferal populations in the bathyal northeast Atlantic: the assemblage response. Deep Sea Res A 37:1263–1283

Lambshead PJD, Brown CJ, Ferrero TJ et al (2002) Latitudinal diversity patterns for deep-sea marine nematodes and organic fluxes—a test from the central equatorial Pacific. Mar Ecol Prog Ser 236:129–135

Lambshead PJD, Brown CJ, Ferrero TJ et al (2003) Biodiversity of nematode assemblages from the region of the Clarion-Clipperton Fracture Zone, an area of commercial mining interest. BMC Ecol 3:1. doi:10.1186/1472-6785-3-1

Lampitt RS (1985) Evidence for the seasonal deposition of detritus to the deep-sea floor and its subsequent resuspension. Deep Sea Res 32:885–897

Lauerman LML, Smoak JM, Shaw TJ et al (1997) ^{234}Th and ^{210}Pb evidence for rapid ingestion of settling particles by mobile epibenthic megafauna in the abyssal NE Pacific. Limnol Oceanog 42:589–595

Levin LA (1991) Interactions between metazoans and large, agglutinating protozoans: implications for the community structure of deep-sea benthos. Amer Zool 31:886–900

Levin LA, Gooday AJ (1992) Possible roles for xenophyophores in deep-sea carbon cycling. In: Rowe GT, Pariente V (eds) Deep-sea food chains and the global carbon cycle. Kluwer Academic Publishers, Amsterdam

Levin LA, DeMaster DJ, McCann LD et al (1986) Effects of giant protozoans (Class: Xenophyophorea) on deep-seamount benthos. Mar Ecol Prog Ser 29:99–104

Loubere P (1998) The impact of seasonality on the benthos as reflected in the assemblages of deep-sea foraminifera. Deep-Sea Res I 45:409–432

Mahatma R (2009) Meiofauna communities of the Pacific Nodule Province: abundance, diversity and community structure. PhD thesis, University of Oldenburg, Oldenburg, Germany

McClain CR, Rex MA, Jabbour R (2005) Deconstructing bathymetric body size patterns in deep-sea gastropods. Mar Ecol Prog Ser 297:181–187

McIntyre AD (1969) Ecology of marine meiobenthos. Biol Rev Cambridge Phil Soc 44:245–290

Menzel L (2011) First descriptions of copepodid stages, sexual dimorphism and intraspecific variability of *Mesocletodes* Sars, 1909 (Copepoda, Harpacticoida, Argestidae), including the description of a new species with broad abyssal distribution. Zookeys 96:39–80

Miljutin DM, Miljutina MA (2009a) Deep-sea nematodes of the family Microlaimidae from the Clarion-Clipperton Fracture Zone (North-Eastern Tropic Pacific), with the descriptions of three new species. Zootaxa 2096:137–172

Miljutin DM, Miljutina MA (2009b) Description of *Bathynema nodinauti* gen. n., sp. n. and four new *Trophomera* species (Nematoda: Benthimermithidae) from the Clarion-Clipperton Fracture Zone (Eastern Tropic Pacific), supplemented with the keys to genera and species. Zootaxa 2096:173–196

Miljutin DM, Gad G, Miljutina MM et al (2010) The state of knowledge on deep-sea nematode taxonomy: how many valid species are known down there? Mar Biodiv 40:143–159

Miljutina MA, Miljutin DM, Mahatma R et al (2010) Deep-sea nematode assemblages of the Clarion-Clipperton Nodule Province (Tropical North-Eastern Pacific). Mar Biodiv 40:1–15

Moens T, Verbeeck L, Vincx M (1999) Feeding biology of a predatory and a facultatively predatory nematode (*Enoploides longispiculosus* and *Adoncholaimus fuscus*). Mar Biol 134:585–593

Mokievsky V, Azovsky A (2002) Re-evaluation of species diversity patterns of free-living marine nematodes. Mar Ecol Prog Ser 238:101–108

Mullineaux LS (1987) Organisms living on manganese nodules and crusts: distribution and abundance at three North Pacific sites. Deep-Sea Res 34:165–184

Mullineaux LS (1988) The role of settlement in structuring a hard-substratum community in the deep sea. J Exp Mar Biol Ecol 120:247–261

Ozturgut E, Anderson GD, Burns RE et al (1978) Deep Ocean mining of manganese nodules in the North Pacific: pre-mining environmental conditions and anticipated mining effects. NOAA Techn Mem, ERL MESA-33

Palmer MA, Gust G (1985) Dispersal of meiofauna in a turbulent tidal creek. J Mar Res 43:179–210

Paul AZ (1976) Deep-sea bottom photographs show that benthic organisms remove sediment cover from manganese nodules. Nature 263:50–51

Pawlowski J, Holzman M (2008) Diversity and geographic distribution of benthic foraminifera: a molecular perspective. Biodiv Conserv 17:317–328

Pearson TH, Rosenberg R (1978) Macrobenthic succession in relation to organic enrichment and pollution of the marine environment. Oceanog Mar Biol Ann Rev 16:229–311

Pfannkuche O (1993) Benthic response to the sedimentation of particulate organic matter at the BIOTRANS station, 47 °N, 20 °W. Deep Sea Res II 40:135–149

Pfannkuche O, Boetius A, Lochte K et al (1999) Responses of deep-sea benthos to sedimentation patterns in the North-East Atlantic in 1992. Deep-Sea Res I 46:573–596

Radziejewska T (2002) Responses of deep-sea meiobenthic communities to sediment disturbance simulating effects of polymetallic nodule mining. Int Rev Hydrobiol 87:459–479

Radziejewska T, Modlitba I (1999) Vertical distribution of meiobenthos in relation to geotechnical properties of deep-sea sediment in the IOM pioneer area (Clarion-Clipperton Fracture Zone, NE Pacific). In: Chung JS, Sharma R (eds), Proceedings of 3rd Ocean Mining Symposium, Goa, India

Radziejewska T, Stoyanova V (2000) Abyssal epibenthic megafauna of the Clarion-Clipperton area (NE Pacific): changes in time and space versus anthropogenic environmental disturbance. Oceanol Stud 29:83–101

Radziejewska T, Drzycimski I, Galtsova VV et al (2001) Changes in genus-level diversity of meiobenthic free-living nematodes (Nematoda) and harpacticoids (Copepoda Harpacticoida) at an abyssal site following experimental sediment disturbance. In: Chung JS, Stoyanova V (eds), Proc Fourth (2001) Ocean Mining Symposium, Szczecin, Poland

Ramirez-Llodra E, Brandt A, Danovaro R et al (2010) Deep, diverse and definitely different: unique attributes of the world's largest ecosystem. Biogeosci 7:2851–2899

Renaud-Mornant J, Gourbault N (1990) Evaluation of abyssal meiobenthos in the eastern central Pacific (Clarion-Clipperton Fracture Zone). Prog Oceanog 24:317–329

Rex MA, Etter RJ (2010) Deep-sea biodiversity: pattern and scale. Harvard Univ Press, Cambridge, Massachusetts

Rex MA, Stuart CT, Hessler RH et al (1993) Global-scale latitudinal patterns of species diversity in the deep-sea benthos. Nature 365:636–639

Rex MA, Stuart CT, Coyne G (2000) Latitudinal gradients of species richness in the deep-sea benthos of the North Atlantic. Proc Natl Acad Sci 97:4082–4085

Rex MA, McClain CR, Johnson N et al (2005) A Source-Sink hypothesis for abyssal biodiversity. Amer Nat 165:163–178

Rice AL, Billett DSM, Fry J et al (1986) Seasonal deposition of phytodetritus to the deep-sea floor. Proc Roy Soc Edinburgh 88B:265–279

Rice AL, Thurston MH, Bett BJ (1994) The IOSDL DEEPSEAS programme: introduction and photographic evidence for the presence and absence of a seasonal input of phytodetritus at contrasting abyssal sites in the northeastern Atlantic. Deep Sea Res I 41:1305–1320

Scharek R, Tupas LM, Karl DM (1999) Diatom fluxes to the deep sea in the oligotrophic North Pacific gyre at Station ALOHA. Mar Ecol Prog Ser 182:55–67

Schratzberger M, Warr K, Rogers SI (2007) Functional diversity of nematode communities in the southwestern North Sea. Mar Environ Res 63:368–389

Seifried S (2004) The importance of a phylogenetic system for the study of deep-sea harpacticoid diversity. Zool Stud 43:435–445

Sharma J, Bluhm BA (2010) Diversity of larger free-living nematodes from macrobenthos (>250 μm) in the Arctic deep-sea Canada basin. Mar Biodiv 41:455–465

Shirayama Y (1984a) The abundance of deep-sea meiobenthos in the western Pacific in relation to environmental factors. Oceanol Acta 7:113–121

Shirayama Y (1984b) Vertical distribution of meiobenthos in the sediment profile in bathyal, abyssal and hadal deep sea systems of the Western Pacific. Oceanol Acta 7:123–129

Shirayama Y, Fukushima T (1995) Comparison of deep-sea sediments and overlying water collected using multiple corer and box corer. J Oceanog 51:75–82

Shirayama Y, Kojima S (1994) Abundance of deep-sea meiobenthos off Sanriku, Northeastern Japan. J Oceanog 50:109–117

Smith CR, Baco AR (2003) The ecology of whale falls at the seep-sea floor. Oceanog Mar Biol Ann Rev 41:311–354

Smith CR, Hoover DJ, Doane SE et al (1996) Phytodetritus at the abyssal seafloor across 10° of latitude in the central equatorial Pacific. Deep-Sea Res II 43:1309–1338

Smith CR, Berelson W, DeMaster DJ et al (1997) Latitudinal variations in benthic processes in the abyssal equatorial Pacific: control by biogenic particle flux. Deep-Sea Res II 44:2295–2317

Snider LJ, Burnett BR, Hessler RR (1984) The composition and distribution of meiofauna and nanobiota in a central North Pacific deep-sea area. Deep-Sea Res 31:1225–1249

Soetaert K, Heip C (1989) The size structure of nematode assemblages along a Mediterranean deep-sea transect. Deep-Sea Res A 36:93–102

Sokov AV, Demidova TA (1992) Ob antitsiklonicheskom vikhre v severo-vostochnoy chasti tropicheskoy zony Tikhogo okeana. Meteorol Gidrol 3:57–64

Soltwedel T (1997) Meiobenthos distribution pattern in the tropical East Atlantic: indication for fractionated sedimentation of organic matter to the sea floor? Mar Biol 129:747–756

Soltwedel T (2000) Metazoan meiobenthos along continental margins: a review. Prog Oceanog 46:59–84

Soltwedel T, Miljutina M, Mokievsky V et al (2003) The meiobenthos of the Molloy Deep (5 600 m), Fram Strait, Arctic Ocean. Vie Milieu 53:1–13

Steyaert M, Garner N, van Gansbeke D et al (1999) Nematode communities from the North Sea: environmental controls on species diversity and vertical distribution within the sediment. J Mar Biol Ass UK 79:253–264

Steyaert M, Moodley L, Nadong T et al (2007) Responses of intertidal nematodes to short-term anoxic events. J Exp Mar Biol Ecol 345:175–184

Tchesunov AV (1997) Marimermithid nematodes: anatomy, position in the nematode system, phylogeny. Zool Zh 76:1283–1299

Thiel H (1975) The size structure of the deep-sea benthos. Int Rev ges Hydrobiol 60:575–606

Thiel H (1983) Meiobenthos and nanobenthos of the deep sea. In: Rowe G (ed) Deep-sea biology. The sea, vol 8. Wiley-Interscience, New York

Thiel H (1991) The requirement for additional research in the assessment of environmental disturbances associated with deep seabed mining. In: Mauchline J, Nemoto T (eds) Marine biology, its accomplishments and future prospect. Hokusen-Sha, Japan

Thiel H (1992) Deep-sea environmental disturbance and recovery potential. Int Rev ges Hydrobiol 77:331–339

Thiel H (2001) Use and protection of the deep sea—an introduction. Deep-Sea Res II 48:3427–3431

Thiel H, Pfannkuche O, Schriever G et al (1988/1989) Phytodetritus on the deep-sea floor in a central oceanic region of the northeast Atlantic. Biol Oceanog 6:203–239

Thiel H, Schriever G, Bussau C et al (1993) Manganese nodule crevice fauna. Deep Sea Res I 40:419–423

Thistle D (1978) Harpacticoid dispersion patterns: implications for deep-sea diversity maintenance. J Mar Res 36:377–397

Thistle D (1979) Harpacticoid copepods and biogenic structures: implications for deep-sea diversity maintenance. In: Livingston RJ (ed) Ecological processes in coastal and marine systems. Plenum, New York

Thistle D (1983) The role of biologically produced habitat heterogeneity in deep-sea diversity maintenance. Deep Sea Res A 30:1235–1245

Thistle D (1998) Harpacticoid copepod diversity at two physically reworked sites in the deep sea. Deep-Sea Res II 45:13–24

Thistle D (2001) Harpacticoid copepods are successful in the soft-bottom deep sea. Hydrobiologia 453(454):225–259

Thistle D, Sedlacek L (2004) Emergent and non-emergent species of harpacticoid copepods can be recognized morphologically. Mar Ecol Prog Ser 266:195–200

Thistle D, Ertman SC, Fauchald K (1991) The fauna of the HEBBLE site: patterns in standing stock and sediment-dynamic effects. Mar Geol 99:413–422

Thistle D, Lambshead PJD, Sherman KM (1995) Nematode tail-shape groups respond to environmental differences in the deep sea. Vie Milieu 45:107–115

Thistle D, Levin LA, Gooday AJ et al (1999) Physical reworking by near-bottom flow alters the metazoan meiofauna of Fieberling Guyot (northeast Pacific). Deep-Sea Res I 46:2041–2052

Thistle D, Sedlacek L, Carman KR et al (2007a) Emergence in the deep sea: evidence from harpacticoid copepods. Deep-Sea Res I 54:1008–1014

Thistle D, Sedlacek L, Carman KR et al (2007b) Exposure to carbon dioxide-rich seawater is stressful for some deep-sea species: an in situ, behavioral study. Mar Ecol Prog Ser 340:9–16

Tietjen JH, Deming JW, Rowe GT et al (1989) Meiobenthos of the Hatteras Abyssal Plain and Puerto Rico Trench: abundance, biomass and association with bacteria and particulate fluxes. Deep-Sea Res 36:1567–1577

Tilot V (1992) La structure des assemblages megabenthiques d'une province à nodules polymétalliques de l'océan Pacifique tropical Est (The structure of megabenthic assemblages in a polymetallic nodule province of the tropical East Pacific Ocean). PhD thesis, Univ Brest, France

Tkatchenko GG, Radziejewska T (1998) Recovery and recolonization processes in the area disturbed by a polymetallic nodule collector simulator. In: Chung JS, Olagnon M, Kim CH et al (eds) Proceedings of 8th ISOPE Conference, vol 2. Montreal, Canada, pp 282–286

Trueblood DD, Ozturgut E, Pilipchuk M et al (1997) The Ecological Impact of the Joint U.S.-Russian Benthic Impact Experiment. In: Proceedings of Second (1997) Ocean Mining Symposium, Seoul, Korea, 24–26 Nov 1997, pp 139–145

Turner JT (2002) Zooplankton fecal pellets, marine snow and sinking phytoplankton blooms. Aquat Microb Ecol 27:57–102

Tyler PA (1988) Seasonality in the deep sea. Oceanog Mar Biol Ann Rev 26:227–258

Tyler PA (1995) Conditions for the existence of life at the deep-sea floor: an update. Oceanog Mar Biol Ann Rev 33:221–244

Tyler PA (2003) Introduction. In: Tyler PA (ed) Ecosystems of the deep oceans. Elsevier, Amsterdam

Vanaverbeke J, Soetaert K, Vincx M (2004) Changes in morphometric characteristics of nematode communities during a spring phytoplankton bloom deposition. Mar Ecol Prog Ser 273:139–146

Vanreusel A, van den Bossche I, Thiermann F (1997) Free-living marine nematodes from hydrothermal sediments: similarities with communities from diverse reduced habitats. Mar Ecol Prog Ser 157:207–219

Vanreusel A, Fonseca G, Danovaro R et al (2010) The contribution of deep-sea macrohabitat heterogeneity to global nematode diversity. Mar Ecol 31:6–20

Varon R, Thistle D (1988) Response of a harpacticoid copepod to a small-scale natural disturbance. J Exp Mar Biol Ecol 118:245–256

Veillette J, Juniper SK, Gooday AJ et al (2007a) Influence of surface texture and microhabitat heterogeneity in structuring nodule faunal communities. Deep-Sea Res I 54:1936–1943

Veillette J, Sarrazin J, Gooday AJ et al (2007b) Ferromanganese nodule fauna in the Tropical North Pacific Ocean: species richness, faunal cover and spatial distribution. Deep-Sea Res I 54:1912–1935

Verlaan PA (1992) Benthic recruitment and manganese crust formation on seamounts. Mar Biol 113:171–174

Verschelde D, Gourbault N, Vincx M (1998) Revision of *Desmodora* with descriptions of new desmodorids (Nematoda) from hydrothermal vents of the Pacific. J Mar Biol Ass UK 78:75–112

Vincx M, Bett GJ, Dinet A et al (1994) Meiobenthos of the deep Northeast Atlantic. Adv Mar Biol 30:1–88

Vopel K, Thiel H (2001) Abyssal nematode assemblages in physically disturbed and adjacent sites of the eastern equatorial Pacific. Deep-Sea Res II 48:3795–3808

Wedding LM, Friedlander AM, Kittinger JN et al (2013) From principles to practice: a spatial approach to systematic conservation planning in the deep sea. Proc Roy Soc Ser B 280: 20131684. dx.doi.org/10.1098/rspb.2013.1684

Wieser W (1953) Die Beziehung zwischen Mundhohlengestalt, Ernahrungsweise und Vorkommen bei freilebenden marinen Nematoden. Ark Zool 4:439–484

Zekely J, Van Dover CL, Nemeschkal HL et al (2006) Hydrothermal vent meiobenthos associated with mytilid mussel aggregations from the Mid-Atlantic Ridge and the East Pacific Rise. Deep Sea Res I 53:1363–1378

Zenkevich LA (ed) (1969) Tikhiy Okean. Biologiya Tikhogo Okeana. Kniga 2. Glubokovodnaya Donnaya Fauna. Pleuston. Nauka, Moskva

Chapter 4
Meiobenthos as a Component of Anthropogenic Disturbance Assessment in the Abyssal Pacific Environment

Abstract Effects of disturbance in the marine environment are assessed, in situ or in the laboratory, based on various indicators, including measures of change in benthic community attributes (abundance, biomass, composition, diversity). The benthic organisms used usually represent the macrobenthos, but the meiobenthos is increasing frequently recommended for such assessments. In anticipation of the polymetallic nodule extraction from the abyssal Pacific areas, a number of field experiments were conducted in which seafloor alteration resembling that accompanying nodule mining, or effects similar to those expected from mining activities, was simulated using various devices: a test miner (the 1975–1980, with a 2006 follow-up, DOMES experiment in CCFZ), a plough-harrow (the 1989–1996 DISCOL experiment in the Peru Basin, S Pacific), and a Benthic Disturber (Benthic Impact Experiments or BIEs: the 1991–1993 US-Russian Joint BIE, the 1994–1997 JET, and the 1995–2000 IOM BIE, all in CCFZ). The severity of impact was assessed by analysing, more or less comprehensively, changes in meiobenthic community-related variables which included qualitative (taxonomic composition, with a finer resolution in nematodes and harpacticoid copepods) and quantitative (abundances of total meiobenthos and of the key taxa, relative abundances) characteristics. Attempts were also made to assess the degree of recovery from the disturbance by re-sampling the disturbed areas at various time intervals post-disturbance. Meiobenthic communities were observed to be affected by the disturbance, reduced abundances immediately post-disturbance being the major community-level manifestation of impact. Effects observed during follow-up studies differed considerably ; although, in most cases, the overall community recovery was recorded, sometimes as early as several months after the disturbance, the composition of both nematode and harpacticoid taxocoenes was altered. The causes underlying the alteration are difficult to be unequivocally explained. The patch mosaic effects which could have been at play could have been accompanied by effects of some natural phenomena such as episodes of phytodetritus sedimentation known to affect deep-sea meiobenthic communities.

© The Author(s) 2014

T. Radziejewska, *Meiobenthos in the Sub-equatorial Pacific Abyss*,
SpringerBriefs in Earth System Sciences, DOI 10.1007/978-3-642-41458-9_4

Keywords Meiobenthos · Deep sea · Pacific · CCFZ · Disturbance · Nodule mining · Impact assessment · Experiment · Free-living Nematoda · Copepoda Harpacticoida

4.1   General Considerations: Meiobenthic Communities in Relation to Environmental Disturbance

Environmental disturbance is, for organisms and communities, a source of stress which can be regarded as a net effect of various processes and events which, taken together, alter conditions of life for organisms and communities and elicit their responses (Boesch and Rosenberg 1981; Pearson 1981). The type and strength of the responses depend on the intensity and persistence of the disturbance on the one hand, and on an organism's, species's, and community's tolerance range and adaptive potential on the other. At an individual level, the responses may vary from reactions in cells and tissues aimed at boosting the immunological resistance to mutations and to physiological changes that might lead to death of the organism affected. If not wiped out by a severe stress, a population responds to disturbance by altering its abundance and biomass, growth, reproduction and feeding rates, sex ratio, mortality rate, and production (Boesch and Rosenberg 1981; Pearson 1981). At the community level, responses to stress/environmental perturbations are observed in, or inferred from, changes in abundance and biomass of the entire community and proportions of its components, reduction of diversity (species richness), a downward shift in the size structure visible as domination of small-sized species, numerical dominance of *r*-selected opportunist species, and elimination of some sensitive ones (Pearson and Rosenberg 1978). Changes in interspecific interactions may be involved as well (Cardinale et al. 2000).

As the seafloor acts as a repository of auto- and allochthonous materials and energy (Adams et al. 1992; Martin et al. 2009), environmental disturbance in marine ecosystems will ultimately affect the sedimentary cover and its benthic communities. If there is no *a priori* knowledge (and very often there is hardly any; Lavering 1994; Warwick 1993), the severity of disturbance effects, hence of the disturbance itself, is inferred from changes in sediment properties and in the benthos as well as from the persistence of those changes in time. The severity of disturbance is judged also by the time it takes for a community to recover, i.e. to revert to its pre-disturbance state and to rebuild its structure, provided such recovery is at all possible.

The literature on environmental stress effects on benthic communities is very extensive (see reviews in, e.g. Gray and Elliott 2009; Magni et al. 2005). Disturbance effects have been, and are being studied for both scientific and utilitarian reasons. The scientific questions addressed include those pertaining to interactions within an ecosystem (ecosystem functioning) and/or to ecosystem robustness or fragility (e.g. Halpern et al. 2007; Nilsson and Grelsson 1995). Studies on disturbance effects are also an important part of research aimed at developing a sound basis for an environmental impact assessment when a planned human activity on the seafloor is perceived as likely to change environmental characteristics of an area of concern (Underwood 1992, 1996).

Usually, elucidation of disturbance effects in benthic communities is attempted by selecting one or more of possible approaches, including field observations and surveys (e.g. Burd 2002), field (e.g. Colangelo et al. 1996) and/or laboratory experiments (Austen and McEvoy 1997; Austen et al. 1998; Widdicombe and Austen 2001), and combinations of those (e.g. Cowie et al. 2000). Non-experimental field surveys are usually performed to collect baseline data with which to describe conditions prevailing prior to the disturbance (if disturbance is predicted to result from a planned activity), to compare conditions in affected and non-affected areas, and to monitor effects induced by the disturbance, including pollution (Burd 2002; Rees et al. 2006).There is again an extensive literature on the design and implementation of field surveys and monitoring activities and on interpretation of the results so that the changes that have occurred and the extent of recovery are perceived and correctly assessed (see, e.g. Green 1979; Hill et al. 2000; Rees et al. 2006; Underwood 1992, 1996 and references therein). Field surveys aimed at accumulating the *a priori* knowledge on an area to be affected are of paramount importance also because the major impediment in assessing the disturbance consequences is the difficulty of disentangling man-made effects from those produced by natural variability (Underwood 1992; Warwick 1993); to quote Lavering (Lavering 1994), "The effects of man-made disturbance are difficult to quantify in settings where natural changes and processes have not been assessed in detail" (p. 201).

Studies on disturbance effects frequently involve experimentally induced small-scale environmental perturbations, directly on the seafloor (field experiments, e.g. Colangelo et al. 1996) or under simulated laboratory conditions (laboratory micro- and mesocosm experiments; e.g. Austen et al. 1998; Cowie et al. 2000).

Effects of disturbance in both field survey- and experiment-oriented research are assessed based on various indicators (Rees et al. 2006). As mentioned above, an array of such indicators includes a measure of changes in a taxon/population/assemblage /community attributes (e.g. Warwick and Clarke 1994), but also on mathematical expressions developed from a combination and transformation of those attributes in the form of various indices (e.g. the marine biotic index, MBI of Borja and Muxika 2005; the index of taxonomic distinctness of Warwick and Clarke 1995; indices of taxonomic and functional relatedness of Warwick and Clarke 2001) as well as other metrics (e.g. a measure of increased variability in the community, Lambshead and Hodda 1994, Warwick and Clarke 1993b) and types of analyses (e.g. meta-analysis, Warwick and Clarke 1993a). In pollution studies, a case is often made for the use of indicator organism or groups of organisms (Warwick 1993), as certain taxa appear to be more sensitive to pollutants/stressors than others, the latter being even capable of benefitting from altered conditions (due to, e.g. reduced competition) and increasing their abundance and/or diversity (Pearson and Rosenberg 1978). Observations of the latter effects have even led to formulation of an explanation, termed the intermediate disturbance hypothesis

(IDH, Connell 1978) which, when applied to the benthic system, postulates that a not too heavy, or not too frequent, disturbance to the seafloor will increase the benthic community's diversity and/or abundance and biomass, in line with the findings of Pearson and Rosenberg (1978). IDH has been tested in a variety of settings with various results; for example, while some authors questioned its validity (e.g. Huxham et al. 2000; a review by Thrush and Dayton 2002), others showed the hypothesis to hold true in a variety of situations (e.g. Barnes 1999; Barnes et al. 2006; Cadotte 2007; Lee et al. 2001; Roxburgh et al. 2004; Thistle 1998; cf. Chap. 3). With respect to the deep sea, Gallucci et al. (2008) suggested that episodic small-scale disturbance might even be necessary for the persistence of rare species, less affected by the disturbance than the more common ones. They referred to the so-called patch-mosaic model in which small-scale patches or the seafloor sediment differ in their properties (e.g. organic matter enrichment, degree of disturbance) and support highly localised successions of species, thus contributing to an overall high diversity of organisms.

Most commonly, effects of seafloor disturbance have been assessed or inferred from studying macrobenthic assemblages (Warwick 1993). However, on account of the traits displayed by the meiobenthos and outlined earlier (cf. Chap. 2), such studies have been increasingly frequently relying on meiofauna. Numerous workers provided evidence of the meiobenthos being a good indicator of environmental changes and of the severity of stress produced by natural causes (natural disturbance) and anthropogenic effects (see a recent review by Balsamo et al. 2012 and references therein).

Warwick (1993) listed the advantages and disadvantages of using the meiobenthos as a benthos component of choice when assessing effects of an impact to the marine environment. The potential disadvantages associated with using the meiobenthos in impact studies involve primarily taxonomic difficulties, which could, however, be mitigated by the sufficiency of higher taxonomic levels (cf. Bett and Narayanaswamy 2014) and cosmopolitan nature of numerous meiobenthic genera. Another disadvantage is the rather poor, compared to that of the macrobenthos, documentation of meiobenthic community responses to disturbance events. On the other hand, the merits of focusing on the meiobenthos when evaluating a disturbance impact seem to be considerable. The advantages of the meiobenthos include non-mobility (and hence responsiveness to local effects) of the organisms, their small size and high densities facilitating quantitative sampling, even from small ships, and the sufficiency of small sediment samples (i.e. ease of transportation and laboratory processing). However, the shortness of generation times means a potentially fast response to disturbance events, but may also mean a faster recovery and hence obliteration of a disturbance effect, particularly when the disturbance is episodic or analysed with a low temporal resolution of sampling.

Natural disturbances impacting meiobenthic communities in shallow and deep waters include hypoxia/anoxia (Powell et al. 1983; Radziejewska and Masłowski 1997); sediment transport induced by strong benthic currents ("benthic storms") (Aller 1997); fast-ice and seaberg scour (Barnes 1999; Lee et al. 2001) as well as

various manifestations of biological activity on the seafloor, e.g. sediment distur-
bance associated with feeding of fish and large invertebrates (Austen et al. 1998;
Creed and Coull 1984; Dye and Lasiak 1986; Schratzberger and Warwick 1998;
Sherman et al. 1983; Warwick et al. 1990), the presence of biogenic structures
(Varon and Thistle 1988), or algal mat effects (Franz and Friedman 2002).

Among anthropogenic perturbations, meiobenthic responses were studied
with respect to disturbance caused by man's direct intervention into the habitat resulting in restructuring of sediment by physical disturbance (Alongi 1985;
Sherman and Coull 1980); fishing (Schratzberger et al. 2002); fish farming (La
Rosa et al. 2001; Mirto et al. 2002); marine mining and dredging (Hobbs 2002;
Jewett et al. 1999; Savage et al. 2001; Schratzberger et al. 2000); oil exploitation
and oil-related pollution (Danovaro 2000; Fleeger et al. 1996), disposal of wastes
(including radioactive wastes) (Pogrebov et al. 1997; Somerfield et al. 1995), CO_2
sequestration on the deep seafloor (Carman et al. 2004).

The latter example is a harbinger for extending the concept of the meiobenthos as a good indicator of stress/disturbance from the coastal/shallow areas to
the deep sea. The subsequent sections focus on such applications involving field
experiments aimed at assessing the extent of human-induced environmental disturbance on the deep seafloor in which the meiobenthic communities were used
as the major disturbance indicator. Ramirez-Llodra et al. (2011) as well as other
workers (e.g. Glover and Smith 2003; Smith et al. 2008) identified extraction of
mineral resources, notably manganese (= polymetallic) nodules, as one of the
major activities that will impact the deep seafloor in the foreseeable future [regard-
less of reservations voiced by Glasby (2002) who expressed doubts as to the value
of polymetallic nodules as a commercially viable source of metals and as to cost-
effectiveness of further investment in deep-sea mining].

Nevertheless, as already indicated in the **Prologue**, in the present deep-sea
R&D climate and given the sequence of claims to securing nodule-bearing areas
within the CCFZ and exploration contracts with ISA (Lodge et al. 2014) as well as
efforts aimed at developing the nodule resources in areas under national jurisdic-
tions (e.g. Fiji, Papua-New Guinea, Cook Islands; Hein and Petersen 2013), there
is a need to address this type of human intervention and to assess its potential consequences using various proxies of impact severity. The subsequent sections will
highlight the past efforts of using meiobenthos as a response descriptor in impact
assessments.

4.2 Deep-Sea Mineral Resources Development *versus* Benthic Communities

As stated by numerous authors (see reviews by Baker et al. 2001 and de
Fontaubert 2001), mineral resource development-related threats to marine environment cannot be analysed without appropriate regard for political, legal, and
technical questions and aspects. Such a far-fetching treatment, however, is far

beyond the scope of this book. The discussion below will focus on potential impacts of polymetallic nodule mining on the abyssal seafloor, as this type of activities and its environmental consequences have received most attention so far.

At present, the literature on potential environmental impacts of seafloor resource exploitation and environmental aspects of deep-sea mining is already quite extensive [e.g. Berge et al. 1991; Chung et al. 2002; Markussen 1994; Thiel et al. 1992]; potential impacts of exploratory activities have been taken into account as well (Thiel et al. 1998). It has been emphasised that potential environmental effects of deep-sea mining will differ in their spatial scales and intensity, depending on a type of resource being exploited (e.g. polymetallic nodules, metalliferous muds, cobalt-rich crusts), location and form of a deposit to be developed, and the volume of material retrieved.

Generally, changes in abyssal communities related to commercial exploitation of nodules can be divided into two categories (Jumars 1981; Markussen 1994). Changes of the first category will take place within the area subjected to disturbance by the mining device (a nodule miner). At the present stage and considering the type of mining device envisaged for use (e.g. Chung 2013; Flentje et al. 2012, Smith and Heydon 2013), the changes will result from destruction and/or disturbance of the surface sediment layer down to the depth of several centimetres. The stress induced there can be expected to be severe, as the entire habitat type (such as nodules) will be eliminated and the remaining part (the soft sediment) considerably physically reworked. The exploitation will always affect—by disturbing or downright destruction—the original structure of the seafloor sediment constituting a habitat of the benthic fauna, morphologically and functionally adapted to inhabiting it (Rhoads 1974). The interference will affect both the organisms dwelling on the sediment surface (the epifauna) and those the mode of life and adaptations of which require them to burrow down to different depths in the sediment (the infauna). The magnitude of hazard for the epifauna is augmented by the fact that numerous epibenthic species are sedentary, permanently attached to hard surfaces provided by, e.g. the nodules (Mullineaux 1987, 1988; Veillette et al. 2007a, b, cf. Fig. 1). When this type of substratum is being exploited—removed from the seafloor, the associated fauna will be completely eliminated. In addition, polymetallic nodule extraction from the surface of or from within the deep-sea sediment is expected to result in an array of interconnected effects. Firstly, there will be direct effects produced by the nodule collector which—should it be towed along the sediment surface, would leave grooves of various depths in the sediment, with some of it turned over and move aside, not unlike an effect of ploughing (Fig. 4.1). It can be expected, as indicated by Jumars (1981) and Thiel et al. (1992), that most benthic organisms will not be able to survive the passage of the collector through their habitat and they will be mechanically destroyed, displaced, and/or sucked out of the sediment.

The second group of changes involves those likely to occur in the sediment at a distance from the actual mining area. The changes would not be directly related to effects of the miner's movement and operation, but would be brought about by displacement and sedimentation (redeposition) of the sediment plume, i.e. a concentration of suspended particulates formed by resuspending the fine sediment fractions in the mining area, which are then transported away from it by near-bed

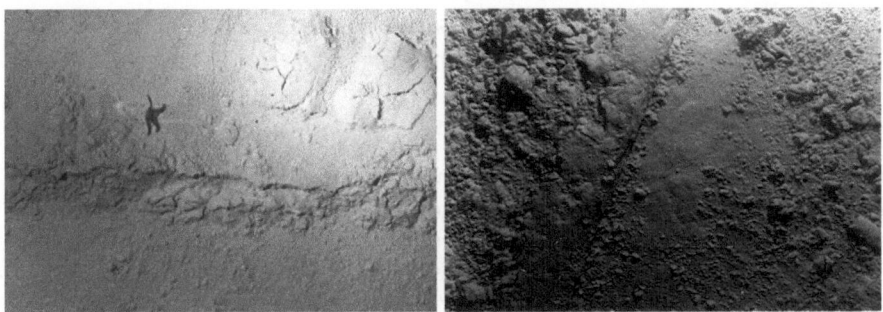

Fig. 4.1 Effects of towing a nodule miner simulator (the Benthic Disturber) on the abyssal sea-floor in the CCFZ during the IOM BIE experiment (*Photos* courtesy of IOM): *left*, a track of one of the Disturber runners; *right*, a Disturber runner track in close-up

currents (Jumars 1981; Thiel and Tiefsee-Unmweltschutz 2001). In this case, the major disturbing effect will be the redeposition of sediment particles from the plume and blanketing of the sediment surface (and organisms inhabiting it) by a layer of particles transported from a distance. The severity of this type of disturbance is likely to be dependent on the mode of plume transport, proximity to the disturbance site, and composition of the particles themselves.

The sediment plume created by resuspension of the surficial sediment or by discharge of mining tailings in the near-bottom water layer will move in different directions, depending on the nodule extraction method, direction and velocity of near-bottom currents, and bottom topography. According to estimates presented by Thiel et al. (1992), at a nodule mining intensity of 1.5 million tonnes a year, a cost-effectiveness threshold [but see Glasby (2000) who estimated this threshold level to be 3 million tonnes annually], about 20,000 tonnes of tailings left after the nodules have been separated will drift above the bottom at a certain distance from the actual nodule mining site. The distance will depend on the suspended particulate size, their ability to aggregate and flocculate, and the sedimentation rate of the emergent aggregates. Effects and ranges of that sedimentation are not known yet, but they ought to be assessed prior to the start of commercial-scale exploitation.

In addition, the collector's operation is likely to disturb the chemical and mechanical equilibrium of the sediment-water interface and of the pore water in the top sediment layer which, as discussed previously, is the major habitat of the infauna, including the meiobenthos. The surficial sediment layer will be disturbed over an area greatly exceeding that directly affected by the moving miner as the lightweight particles of the very finely-grained sediment will be resuspended, while their redeposition—not necessarily where they have been resuspended from—may blanket the epifauna along the plume resedimentation trajectory (Jumars 1981; Thiel et al. 1992).

To enable the development of a predictive capacity towards assessing likely environmental consequences of nodule mining, a number of research programmes were already designed and implemented. They will be discussed in the subsequent sections.

4.3 Programmes Aimed at Assessing Mining Effects on Abyssal Benthic Communities

4.3.1 Background Considerations

In view of the hazards to the seafloor environment and the communities inhabiting it, likely to be produced by abyssal resource exploitation, it is imperative that mining operations, when they commence, be carried out so that the undesirable consequences are minimised. This can be achieved through selection of appropriate mining technologies and equipment. Therefore, methods and approaches have been sought with which to assess the magnitude and scope of the hazards and to estimate rates of habitat recovery and recolonisation by the benthos (Thiel 2001).

The first step in this direction is the evaluation of types, intensity, and magnitude of potential harmful effects and the rate with which benthic communities could revert to a state resembling the original one, once the disturbing activity has been discontinued, i.e. to assess a potential for possible recovery and recolonisation of the disturbed seafloor. One of the approaches to collect this kind of information is to conduct field experiments involving disturbance to the natural environment of the abyssal seafloor. Despite the poor knowledge on the natural state of benthic communities (including, and particularly, the meiobenthos) to be impacted and natural variability experienced by those communities and their habitat, such approach has proven quite popular and a series of experiments were conducted, as discussed below.

4.3.2 DOMES (Deep Ocean Mining Environmental Study), USA, 1975–1980

The Project DOMES was initiated by the National Oceanic and Atmospheric Administration (NOAA) of the United States to collect data with which to estimate and predict environmental effects of polymetallic nodule mining from the abyssal seafloor of the NE Pacific (Ozturgut et al. 1981). The project was carried out in two phases, termed DOMES I and DOMES II. In the first 3 years (DOMES I), the research was aimed at acquiring information on oceanographic variability of the area to be test-mined and at assessing the level of biotic variables in the nodule mining area at 4 selected sites (Fig. 4.2). The scientific part of DOMES I included studies on surface and near-bottom currents, nutrient and suspended particulate profiles, water column acoustics and optics, distribution of phyto- and zooplankton, bottom sediments, nodules, and benthos. The results were summarised in a collection of papers edited by Bischoff and Piper (1979 in Ozturgut et al. 1981). In addition, a comprehensive treatment of DOMES results can be found in Morgan et al. (1999). The benthic research of DOMES I [Hecker and Paul (1979) in Grassle and Morse-Porteous (1987)] targeted the macrobenthos which was analysed based on the contents of 80 box corers from three DOMES sites. Although

Fig. 4.2 Sites of experiments
aimed at assessing potential
impact of nodule mining in
the Pacific abyssal seafloor:
1, DOMES (*left* to *right*: sites
A, B, C, D); *2*, DISCOL; *3*,
US-Russian BIE; *4*, JET; *5*,
IOM BIE

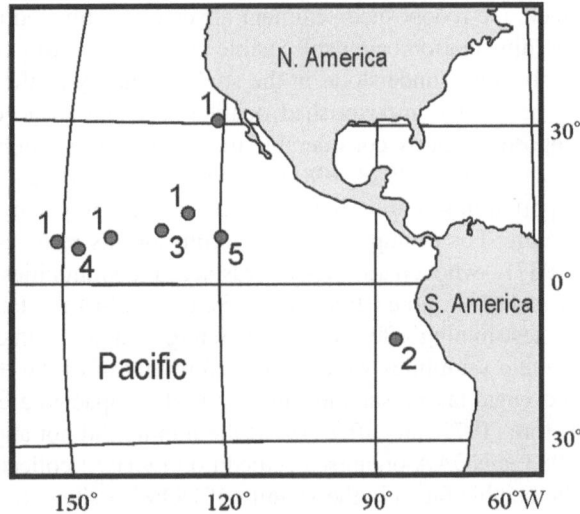

the mesh size used (300 μm) was too coarse for a quantitative estimate of meiofaunal abundance, Hecker and Paul (1979 in Grassle and Morse-Porteous 1987)
reported the meiobenthos to account for 62 % of the total metazoan fauna. During
the subsequent stage (DOMES II) in 1978, two mining tests of engineering fea-
sibility were conducted by the US companies Ocean Management, Inc. (OMI; at
site A of Fig. 4.2) and Ocean Mining Associates (OMA; at site C of Fig. 4.2), and
effects of this small-scale test mining were monitored. In both tests, the nodules
were retrieved by sled-mounted collectors (about 5 m long, 3 m wide, 2 m high)
towed on the bottom, attached (with a flexible pipe) to a rigid drill pipe serving as
a nodule conduit to the mining vessel (Chung 2013). The collectors were designed
so that they removed nodules from the seafloor, winnowed much of the fine sediment, rejected large nodules, and transferred the remaining nodules to the lifting
pipe intake (Ozturgut et al. 1981). Therefore, the mining operation involved creation of sled grooves (tracks) in the sediment, removal of nodules, and generation
of a sediment cloud (plume) in the water column and on its surface, due to surface
discharge of the mining effluent. In both tests, hundreds of metric tonnes of nodules were brought on board the mining vessels. The tests were of a short duration,
and the subsequent assessment, as described by Ozturgut et al. (1981), addressed
the short-term effects as well.

Results of effects of mining tests run by OMI on the benthos were summarised by Ozturgut et al. (1981) based on detailed description by Burns et al. (1980
in Ozturgut et al. 1981). The plume of suspended material over the bottom was
several metres thick and was visible at the distance of 16 km from its origin for
6 days following its formation. Photographs of the seafloor showed a very substantial disturbance of the seafloor surface by the nodule collecting device used. The
collector pushed the bottom material aside, on the outside of the sled tracks, with
heavy redeposition occurring near the tracks. Five to 10 m away from the collector

track, the redeposited sediment cover was estimated to be a few centimetres thick, resedimentation being still visible up to 100 m away on either side of the track. The zoobenthos (understood in the study primarily as the macrobenthos) was expected to be greatly impoverished qualitatively and reduced quantitatively. However, as opposed to fairly considerable treatment of test mining effects on the pelagic biota (Ozturgut et al. 1978, 1981), no post-mining test data on the benthos have ever been reported in scientific literature. There exist only some statements in the "gray" literature. For example, when reporting on his research related to DOMES, Wilson (1987)—who characterised crustacean communities of the DOMES area (sites A and C)—quoted Jumars and Self (unpubl.) as stating that site A was impacted by test mining effective in generating a large sedimentary plume at the sea floor, benthic samplings were made before and after the event, navigational difficulties prevented taking samples directly in the impacted areas, and the samples collected before (1977) and after (1978) the mining did not appear significantly different. In 1983, a NOAA-organised Expedition ECHO I collected 15 quantitative samples of the benthic fauna in the vicinity of DOMES site C (Anonymus 1987; Wilson 1987) to study potential impacts on the benthic community of the OMA test mining conducted 5 years earlier. As reported by Anonymus (1987), the fauna (macrobenthos) from the immediate test mining area was compared with the fauna from an area far enough away to be regarded as undisturbed. It was concluded that the disturbance to the seafloor was either not extensive enough to produce a statistically detectable difference in the community structure between the two areas or the impacted site recovered within 5 years separating the DOMES and ECHO I campaigns.

Despite the limited scope of the DOMES biological research, the project resulted in recommendations (Ozturgut et al. 1981; National Research Council 1984) put forth for the future research to address the following issues:

- direct effects of mining on the benthos within the nodule collector grooves (destructive effects of the nodule collector itself);
- consequences of mining activities for the benthos away from the nodule collector operation (plume sedimentation effects).

These points were taken by subsequent studies, both in the DOMES area and in abyssal nodule areas elsewhere.

The DOMES area (OMA site C, at present within the French mining claim; cf. Figs. 3 and 4.2) was revisited in 2004, 26 years post disturbance, by the French expedition NODINAUT on board RV *L'Atalante*. Observations of and measurements at the seafloor were made using the submersible *Nautile* and benthic landers. Sediment samples were collected to assess the recovery rate in terms of geochemical parameters of the sediment (Khripounoff et al. 2006) and the meiobenthos (Mahatma 2009; Miljutin et al. 2011). The study focused on a single track left by the mining device used to produce disturbance during the DOMES experiment, still distinctly visible on the seafloor 26 years post-disturbance (Khripounoff et al. 2006; cf. Fig. 1.1 in Miljutin et al. 2011). Khripounoff et al. (2006) estimated the track to have been 4.5 cm deep and the incision produced by the experimental device to have remained visible on the sediment surface. However, despite the persistence of the altered sediment surface, Khripounoff et al. (2006) found nutrient (nitrate, silicate, phosphate) profiles not to differ between the formerly disturbed area and the undisturbed seabed in the vicinity. In addition, the oxygen profiles and oxygen consumption in the sediment (a measure of the organic carbon mineralisation, i.e. benthic metabolic rate) in the former track proved not to differ from the adjacent area. This has prompted Khripounoff et al. (2006) to conclude that, after 26 years, the test-mining disturbance effects were not perceptible with respect to the sediment-dwelling fauna. The authors cited went on to state that "After 26 years, this benthic fauna seems to have completely recolonized the track sediment" (p. 2039).

The meiofaunal responses inferred based on the results of the 2004 NODINAUT expedition were addressed in detail by Mahatma (2009) and Miljutin et al. (2011). While Mahatma (2009) focused on meiobenthic communities as a whole and harpacticoid copepods in particular, Miljutin et al. (2011) analysed nematode assemblages. Mahatma (2009) found no difference between the formerly impacted and non-impacted areas in terms of the mean total meiobenthos densities (55.35 $\pm$ 25.51 and 42.32 $\pm$ 20.98 inds/10 cm^2, respectively), mean nematode densities (44.34 $\pm$ 20.82 and 35.67 $\pm$ 17.49 inds/10 cm^2, respectively), and mean copepod (including copepodites) densities (9.97 $\pm$ 5.13 and 5.54 $\pm$ 3.23 inds/10 cm^2, respectively). In terms of composition of the harpacticoid assemblage, the only conspicuous difference reported by Mahatma (2009) was a higher proportion of representatives of the harpacticoid families Zosimidae and Canthocamptidae in the impacted area (10 and 11 %, respectively) than in the intact sediment (5 and 6 %, respectively). Thus the analyses reported by Mahatma (2009) lend support to conclusions of Khripounoff et al. (2006) of a benthic community recovery having taken place in the impacted area despite the still visible external signs of impact on the seafloor surface.

However, these conclusions proved in sharp contrast with those expounded in a later publication of Miljutin et al. (2011) who provided an account of analyses conducted on the same material from the same set of sites. Having analysed the composition and abundance of nematode assemblages in the track-affected and intact (nodule-bearing and nodule-free) bottom areas, Miljutin et al. (2011) reported differences between assemblages in the disturbed and non-disturbed habitats. The differences involved a significant change in the mean nematode abundance (30.7 inds/10 cm^2 in the affected area versus 69.5 and 136.9 inds/10 cm^2 on the unaffected bottom in its nodule-bearing and nodule-free sediment, respectively). There was also a difference in taxonomic composition; analyses showed *Oncholaimus*, *Thalassomonhystera*, and *Acantholaimus* to be the genera most conspicuously contributing to the assemblage-level difference: the track-affected bottom showed a higher abundance of the first two genera and a lower abundance of the third genus compared to the values found in the intact sediment. Miljutin et al. (2011) who identified their nematodes to morphospecies, reported also the presence of differences at the sub-genus level. Analyses involving the diversity indices at various taxonomic levels (family, genus, species), including rarefaction, showed the taxon richness at the formerly disturbed site to be significantly lower than in the unimpacted sediment.

Thus, based on the Control-Impact portion of the BACI (Before-After; Control-Impact) approach (Smith 2002), the major conclusions from the analyses performed by Miljutin et al. (2011) are that the differences in the nematode assemblage induced by the experimental seafloor disturbance are visible and persistent, and that the disturbance impact is clearly discernible using a proxy such as a nematode assemblage, a component of the meiobenthos in the deep sea.

4.3.3 DISCOL (Disturbance and Re-Colonization Experiment in a Manganese Nodule Area of the Deep South Pacific), Germany, 1989–1996

DISCOL and the associated follow-up programmes (Schriever 1995; Thiel and Forschungsverbund Tiefsee-Unmweltschutz 1995, 2001) were carried out by German researchers in the Peru Basin of SE Pacific (Fig. 4.2) at depths of about 4,150 m. The first stage of DISCOL (in 1989) involved an experimental disturbance of about 11 km^2 of the seafloor (the DISCOL Experimental Area, DEA) effected with a specially constructed device termed the plough-harrow (Schriever 1995). The disturbance produced by the device involved ploughing the sediment along predetermined circular trajectories. Instead of simulating an actual nodule mining operation, the use of the plough-harrow was intended to physically disturb the sediment in a way similar to what could be expected from the caterpillar tracks or chains of a self-propelled nodule collector carrier: resuspension of the surficial sediment layer and sediment redeposition. The disturbance produced a mixture of very soft surface sediments and more consolidated deeper deposits. Only a small amount of fine sediment was resuspended and resettled. The disturbance was unevenly applied to the DEA in that the heaviest impact occurred at the centre, the peripheral circles being much less affected (Schriever et al. 1991). For sampling purposes, the DEA was divided into 8 wedge-shaped sectors positioned along the main geographic directions, and each of the wedges was additionally subdivided into 3 zones of impact intensity diminishing gradually from the centre towards peripheries (Schriever et al. 1997).

Disturbance effects were monitored immediately after the fact as well as after half a year, 3 and 7 years. Three years after the disturbance, the plough-harrow track contours had lost their sharp-edged appearance, but remained recognizable when viewed 7 years post-disturbance (Borowski 2001; Thiel and Tiefsee-Unmweltschutz 2001).

DISCOL and its follow-up programmes were notable in that the metazoan meiobenthos featured prominently on their research agendas, as it was selected—in addition to the megafauna and macrobenthos—as one of the three biological proxies to be used in the assessment of the severity of disturbance and the subsequent recovery. The meiobenthos-related variables used by the DISCOL researchers included composition of meiobenthos assemblages, their total and relative abundances as well as the abundance and taxonomic composition of the major higher taxa—nematodes and harpacticoid copepods. An important consideration in the sampling design from the standpoint of reliability and comparability of quantitative data was the use of the multiple corer (multicorer) to collect samples for the quantitative study of meiofauna. Comparisons were made between data from multicorers deployed in plough-harrow tracks and between them. The post-impact study involved also sampling an unaffected reference site 3 Nm south of the DEA.

Immediately after the disturbance, the total metazoan meiobenthos abundance was observed to be reduced, the reduction being maintained until the follow-up sampling half a year after disturbance (Schriever et al. 1997; Thiel and Forschungsverbund Tiefsee-Umweltschutz 2001). Particularly affected were the harpacticoids. Three years later, however, the abundances were higher than the initial values: the densities of the Nematoda had increased to levels more than twice those of the pre-impact study, abundances of harpacticoids increasing by more than 60 %. Incidentally, the abundances in the control—unimpacted—area were higher as well, which could indicate a general longer-term variability of which no information had been initially known (Schriever et al. 1997). The samples collected 7 years after the sediment had been ploughed showed the overall meiobenthic densities to have decreased, but the abundances of both nematodes and harpacticoids were by 50 and 15 %, respectively, higher than the respective pre-disturbance levels, while the data from the reference area showed values almost identical to the pre-disturbance ones (Schriever et al. 1997). When sampled 7 year post-disturbance, the meiobenthic assemblages from the previously impacted and unimpacted seafloor areas showed differences in abundance and composition of nematodes and harpacticoids: the nematode mean absolute and relative abundances were higher in the previously disturbed area (163.47 ± 33.08 inds/10 cm^2; 79.66 %, respectively) than in the undisturbed and reference parts of the seafloor ($109–123.29 \pm 3$ 1.80 inds/10 cm^2; 71.74–77.07 %, respectively), a reverse being the case in the relative abundance of harpacticoids ($17.66–21.45 \pm 2.61$ inds/10 cm^2 and 12.5 % in undisturbed and reference areas versus 17.81 ± 2.55 inds/10 cm^2 and 8.68 % in the previously ploughed seafloor), but those differences were not statistically significant (Ahnert and Schriever 2001).

Vopel and Thiel (2001) studied nematodes in samples collected during the 1996 cruise (7 years post-disturbance) and addressed differences in the nematode taxocoene between the formerly impacted and non-impacted areas by randomly selecting 5 cores from each. They found nematodes to account for 77.2 and 79.3 % of all meiobenthic metazoans in their samples from unimpacted and impacted sites, respectively. Analyses brought down to the genus level revealed *Acantholaimus* to be dominant (about 20 % in both sets of samples), *Thalassomonhystera* being the genus discriminating between the site types (7.4 and 14.2 % in the control and disturbed sites, respectively). Univariate tests failed to detect significant differences in genus-level diversity, evenness, and richness values between the two site types. On the other hand, multivariate analyses Vopel and Thiel (2001) performed showed no significant difference at the genus level, but the differences between the two site types turned out to be significant at the family level, the Monhysteridae and Aegioloalaimidae being the discriminant families (contributing the most to the dissimilarity between the two sets of data). Vopel and Thiel (2001) explained the absence *versus* the presence of differences between the impacted and control sites by referring to a higher taxonomic variability at the genus level in the undisturbed samples, whereas aggregation to the family level eliminated this source of variability. The family Aegialoalaimidae was absent in samples from the control sites, while the monhysterids were abundantly represented in both types of samples on account of the abundance of the genus *Thalassomonhystera*, dominant in the nematode taxocoene. The mean relative abundance of the genus, however, in the disturbed area was twice that at the control sites. When presenting their results, Vopel and Thiel (2001) admitted the lack of adequate knowledge with which to explain both the pattern observed and the responses of abyssal nematode populations to changes in their physical, biotic, and geochemical environments.

The harpacticoid taxocoene in the DISCOL study was analysed at the family level for all stages of the study (Ahnert and Schriever 2001). Comparison of the pre-impact with immediate and later post-impact results revealed no differences. However, when monitored 7 years post-disturbance, a significant difference in harpacticoid abundance between the undisturbed and disturbed sites was revealed, the Argestidae, Ameiridae, and Thalestridae turning out to be most important in discriminating between the two types of habitats. Commenting on the utility of harpacticoids in assessing the impact, i.e. discriminating between the impacted and non-impacted sites, Ahnert and Schriever (2001) referred to the traits indicator organisms should possess. The most important traits include an ability to distinctly respond to a disturbance as well as the abundance, large size, and ease of identification. With respect to the first criterion, the discriminant families identified in their study (argestids, ameirids, and thalestrids) were all contributing most consistently (with lowest variation) to the significance of differences. The second criterion (a combination of abundance, size, and ease of identification) was distinctly satisfied primarily by argestids and ameirids (although not so much by thalestrids).

Based on the results of DISCOL and associated programmes, Schriever et al. (1997) concluded that, despite limitations revealed, the meiobenthos proved a good proxy for studying the disturbance impact. Nematodes (at the genus level) and harpacticoids (at the family level) met the requirements of disturbance and recovery indicator taxa and could be recommended for use in other mining impact and recolonisation studies.

4.3.4 The "Benthic Disturber": A Nodule Mining Simulator

In the early 1990s, on the wave of intensified exploration efforts in the Clarion-Clipperton nodule fields, the National Oceanic and Atmospheric Administration (NOAA) of the United States initiated a series of field experiments involving disturbance to the seafloor that would mimic the actual nodule mining. In all these experiments, the disturbance was created using a device designed and constructed in 1991 by the Sound Ocean Systems, Inc. (SOSI), a Seattle (USA)-based

Fig. 4.3 The Benthic Disturber on board RV *Yuzhmorgeologiya* (*Photo* courtesy of IOM)

company, and termed the Deep Sea Sediment Resuspension System (DSSRS)
(Fig. 4.3; Barnett and Yamauchi 1995, Brockett and Richards 1994). The device,
commonly referred to as the Benthic Disturber (or simply the Disturber), is a
passively towed two-runner sled-type vehicle designed to simulate the sediment

disturbance produced by a nodule collector. The sled is 4.8 m long, 2.4 m wide, and 5.0 m high. The device is equipped with two pumps: when the Disturber is being towed on the seafloor, one of the pumps fluidises (loosens) the surficial sediment, while the other assists in lifting the sediment (possibly with nodules) and feeding it into a 300 mm vertical riser to eventually discharge it into the water column about 5 m above the seabed. An electronically activated water sampler system consisting of a rosette with 12 Niskin bottles can be mounted on the riser to collect samples of the discharged slurry and determine the amount of resuspended sediment. The Disturber is a tethered device attached to a tow cable at the front of the sled, and is controlled by telemetry system signals transmitted from the control room on board a research vessel via a coaxial cable. The cable transmits also information from sensors (compass, altimeter, and pitch and roll sensors) mounted on the device to the onboard controller; this information makes it possible to adjust the position, orientation, and operation of the system. The system is operated from an onboard computer; the sensor information and images from a video camera mounted at the top of the discharge shaft are received in the real time and stored in the computer hard disk for further post-operation processing. In addition, the Disturber is equipped with acoustic responders mounted on the depressor (attached to the tow cable to provide a favourable towing angle); the responders communicate with the vessel's underwater navigation system consisting of four acoustic transponders deployed within the test site. This way, the position of the Disturber (or, more accurately, of the depressor) relative to the ship and various observation moorings placed in the vicinity of the tow zone can be monitored and recorded.

4.3.5 Joint US-Russian BIE (Benthic Impact Experiment), USA-Russia, 1991–1993

The first of a series of field experiments (called BIEs, or Benthic Impact Experiments) involving the Disturber was carried out in 1991–1993 on the initiative of NOAA. It was a joint American-Russian endeavour conducted at a test site in the Russian claim area within the central part of the Clarion-Clipperton nodule field (Fig. 4.2) at the depth of about 4,800–4,900 m. Initially (1991–1992), the newly built first-generation Disturber was tested and experimented with. As the tests were not satisfactory, the second-generation Disturber was constructed, tested in 1993, and deemed satisfactory (Brockett and Richards 1994) so the actual experiment could be carried out. The baseline study in July 1993 involved assessment of current measurements, sonar and camera surveys, and sediment box coring. The meiobenthos was analysed based on box corer subsamples. The actual experiment was run in August 1993, whereby the Disturber was towed along the bottom 49 times, incising the sediment down to about 29 mm (as estimated by Yamazaki and Sharma 2001), and resuspended an estimated total of 4,000–4,328 m^3 of sediment in the process (Trueblood and Ozturgut 1997; Trueblood et al. 1997; Yamazaki and Sharma 2001). Nine months after the disturbance (in 1994), the site was revisited to sample the macro- and meiobenthos again, the meiobenthos being sampled with a multicorer. The main thrust in the assessment of towing impacts was directed to effects of resedimentation, no within-track sampling being performed. Moreover, the analysis of benthic response addressed primarily the macrobenthos, and detailed analyses of macrobenthic responses, particularly of the polychaetes (at the family level), were published by Trueblood and Ozturgut (1997) and Trueblood et al. (1997). Although the overall macrobenthic abundance remained unaffected by the redeposited sediment layer (estimated to be only 1 mm thick), two macrobenthic taxa, the polychaete family Sabellidae and the isopod family Macrostylidae, were found to exhibit a significant response to resedimentation by increasing their abundances in the affected area. In addition, a similar response was evident in the glycerid polychaete *Glycera* sp., known to be a scavenger and probably responding to the likely increase in food resources provided by the redeposited sediment.

Results presented by Trueblood and Ozturgut (1997) and Trueblood et al. (1997) for the meiobenthos concerned a comparison of treatment (resedimentation) and control samples collected immediately after the disturbance in 1993 and showed no significant difference in the nematode abundance. The abundance recorded 9 months later was, however, significantly lower in the disturbed treatment (resedimentation) zone. In contrast, harpacticoid copepod abundances showed no effect immediately after the disturbance and during the follow-up cruise.

Nematodes in some selected samples (picked up from those collected in the control area in 1993 and, in 1994, in both the control area and that previously impacted by resedimentation) down to 2 cm in the sediment were identified to the genus level by V. Galtsova and L. Kulangieva (pers. comm.), and harpacticoid copepods from still a lower number of samples were similarly identified by I. Drzycimski (pers. comm.). No published information on those identification is available yet. However, as reported by Radziejewska et al. in an unpublished conference paper (see footnote to Table 3.1 for details), the nematode assemblage structure showed no marked between-treatment effect (1994 data), but the structure did differ, in terms of genus richness and genus-based diversity indices, between the 2 years of study, perhaps reflecting the difference in sampling gear employed (a box corer in 1993 and a multicorer in 1994). The low number of samples analysed for harpacticoid copepods precluded any meaningful statistical treatment of data.

The Joint US-Russian BIE site has not been resampled since 1994, so no information is available on the course of recovery and recolonisation processes there.

4.3.6   JET (Japan Deep-Sea Impact Experiment), Japan, 1994–1997

Japan was another ISA contractor to have carried out a BIE-type experiment in the Clarion-Clipperton nodule field. The experiment was designed to address a number of questions concerning the magnitude of impacts and impact mechanisms, to be inferred from responses of the meiobenthos deemed the faunal category most appropriate for providing the desired answers and insights (Fukushima 1995). The first question was whether artificially generated rapid sediment deposition affects the density, vertical distribution, and composition of the meiobenthos; it was expected that this question would be answered by data on properties of meiobenthic communities in relation to the redeposited sediment thickness, a measure of impact severity. The other question was addressed by assuming that the reredeposition changes (dilutes) the food resources of the meiobenthos, hence the answer was to be sought by measuring the density of sediment bacteria, assumed as a proxy of meiobenthic food resources. The experiment, run in the Japanese claim area in the western part of the CCFZ (cf. Figs. 3 and 4.2), began after a baseline survey (JET1) conducted in 1994 during which water column and sediment properties were determined, and sediment samples were collected for the assessment of initial meiobenthos density, composition, and distribution and for the estimation of bacterial biomass. The actual experiment, conducted immediately after the baseline survey (JET2), involved creation of sediment disturbance with the Benthic Disturber described above. The 19 completed Disturber tows resulted in resuspension of an estimated total of about 2495 m^3 of sediment (Fukushima 1995; Yamazaki and Sharma 2001), the depth of incision into the sediment being estimated at about 50 mm (Yamazaki and Kajitani 1999), subsequently re-estimated at exceeding 44 mm (Yamazaki and Sharma 2001). Sediment samples were again collected to estimate the parameters of meiobenthic assemblages selected. Incidentally, in his account of the two first JET cruises (JET1 and JET2), Fukushima (1995) reported difficulties in using the multicorer to collect sediment from the seafloor densely covered with nodules, for which reason the number of intact cores was in both cases much lower than planned. The JET site was resampled in two subsequent campaigns, JET3 (in 1995, about 1 year post-disturbance) and JET4 (in 1996, about 2 years following the disturbance) (Yamazaki and Kajitani 1999).

When analysing the biological data of JET, distinction was made between zones of resedimentation intensity (low, medium, and high), as estimated from video surveys (Shirayama 1999) and model calculations (Taguchi et al. 1995). Immediately after the disturbance (JET2 cruise), the meiobenthos showed a dramatic (and significant) decline in density (from an average of about 105.9 to about 47.9 inds/10 cm^2), overall and in each sediment layer (down to 3.0 cm) analysed. Interestingly, the decline was not related to the severity of resedimentation. A similarly pronounced and significant decline was observed in the density of nematodes, whereas neither the copepods nor the sediment bacteria cell counts—despite highly reduced mean abundance values—proved significantly affected [Kaneko (Sato) et al. 1995, 1997; Shirayama 1999]. At the subsequent stages of JET (JET3 and JET4), the meiobenthic communities were assumed to have recovered, as concluded from increased total abundances in the previously impacted area regardless of the initial resedimentation severity [Kaneko (Sato) et al. 1997; Shirayama 1999]. A similar pattern was displayed by nematodes [which is understandable, considering that nematodes made up most of the meiobenthos, in excess of 90 % in JET1 and more than 80 % in JET2; Kaneko (Sato) et al. 1995]. However, Shirayama (1999) referred to his

unpublished data on nematode taxonomic composition to caution against the conclusion of a complete recovery (despite the increased abundance), as the taxonomic
composition of the nematode taxocoene in JET4 proved completely different from
that existing during JET1. On the other hand, the harpacticoid copepod abundances
observed in the follow-up cruises were observed to rise considerably regardless of
the resedimentation regime (and in the control area) during JET3 [Kaneko (Sato)
et al. 1997; Shirayama 1999], to decrease again (but to levels higher than those at
the pre-disturbance stage) in JET4 (Shirayama 1999). Bacterial counts in JET3 were
observed to soar high above the levels recorded in both JET1 and JET2 [Kaneko
(Sato) et al. 1997], no data being reported for JET4.

Regarding changes in the vertical distribution of meiobenthos, Shirayama (1999)
reported a drastic change in the low-resedimentation area where abundances lev-
elled off downcore contrasting with the exponential decline in densities with sediment depth prior to the disturbance. Interestingly, no changes in vertical distribution
were observed in the medium- and high-resedimentation area, although the densities
decreased in every layer analysed, suggesting the impact reaching deeper in the sediment at a more intensive habitat alteration by sediment redeposition.

It is tempting, given the meiobenthic focus and frequency of sampling after the
disturbance, to assess to what extent the question posed initially were answered by
the JET results. It may be contended that, in this case, the meiobenthos in general,
and nematodes in particular, proved a good proxy for the assessment of initial effects
of disturbance. The first question addressing a possibility that artificial rapid sediment deposition affects meiobenthis parameters, can be answered in the affirmative.
The effects were clearly discernible in the abundances and vertical distribution of
the meiofauna and its major sediment-bound component, the nematodes. However,
the relationship with the severity of disturbance (thickness of the redeposited sediment layer) proved unclear, most probably on account of a generally very thin layer
of resedimented particles. On the other hand, the second question proved difficult
to answer, although enhanced bacterial densities revealed during the follow-up
cruise (JET3) could speak for increased trophic resources facilitating recovery of the
meiobenthic community. It may be speculated that the enhanced bacterial densities
could be, at least in part, responsible for the alteration of the taxonomic composition
of the nematode taxocoene referred to by Shirayama (1999), should the taxocoene's
dominance structure be found to have switched to domination of selective deposit
feeders (cf. Chap. 3, Fig. 3.1) adapted to ingestion of bacterial-sized food particles.
However, there are no published data to support, or otherwise, this speculation.

4.3.7   IOM BIE (Interoceanmetal Benthic Impact Experiment), 1995–2000

IOM BIE was conceived as a complement to the set of Benthic Disturber-
employing BIE-type experiments carried out in the Clarion-Clipperton nodule field (Kotlinski and Stoyanova 1998). The experiment was carried out, in

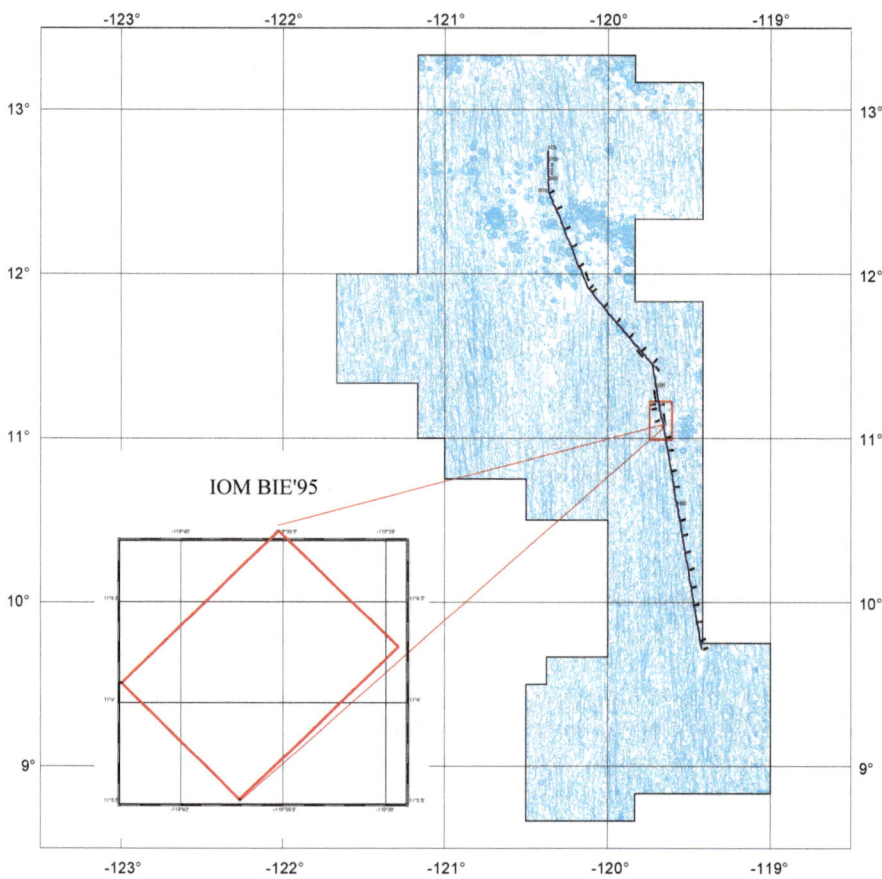

Fig. 4.4 Location of the IOM BIE test site on a baseline survey transect within the IOM claim area's Sector B1 (drawing courtesy of R. Kotliński)

July 1995, within a 1.5 × 2 km test site situated in a nodule-free patch (cf. Fig. 2.5) surrounded by an extensive nodule field within Sector B1 of the IOM claim area (eastern part of CCFZ) (Fig. 4.4).

Selecting a nodule-free patch for the test site had been a conscious decision made to facilitate sediment sampling and to avoid sampling difficulties encountered during JET (see above; A. Pařizek, pers. comm.), but this created a somewhat artificial situation with respect to simulation of nodule mining. As a consequence of the test site choice, rather than mimicking nodule sampling, the test involved creation of a general sediment disturbance the effects of which were subsequently evaluated.

The test itself was preceded by a baseline study, began in 1994, aimed at obtaining a picture of reference oceanographic, geological and biological conditions, the relevant data being collected from the 330-km long survey transect mentioned above (Fig. 4.4; Kotlinski and Stoyanova 1998). Results of this survey were also used to select the actual test site. While designing the test, it was intended to create disturbance to the seafloor by repeatedly towing the Benthic Disturber (described in Sect. 4.3.4 above) diagonally within the site selected, and to assess the extent of disturbance using the meiobenthos-related metrics (total abundance, vertical distribution, abundance and taxonomic composition of nematodes and harpacticoids identified to the genus level) as the response variables (Radziejewska 2002). The initial status of sediment characteristics, including pore water chemistry, was determined as well. The test itself was conducted in July 1995, following the collection of pre-impact sediment cores to be examined for the initial status of the response variables. The disturbance was effected by a total of 14 Disturber tows which resuspended an estimated 1,800 m^3 of sediment. Immediately after the towing was completed, the test area was photo-video-surveyed and the sediment was sampled to provide data on the meiobenthic response variables as well as on the accompanying environmental parameters. Photo-surveys demonstrated the sediment surface in the tow track to be severely, although not uniformly, mechanically altered (cf. Fig. 4.1); the photographs showed Disturber runner tracks visible as sharp-edged furrows, with the surficial sediment layer pushed aside and overturned (Tkatchenko and Radziejewska 1998). The video footage provided indication of an increased feeding activity of the motile epifauna (fish and shrimps; Radziejewska, pers. obs.), suggesting a release of easily available food resources, probably in the form of dead infauna, by the disturbance.

Twenty two months after the disturbance, in April 1997, the test site was revisited (the IOM BIE-97 cruise), the subsequent revisit taking place in June 2000 (IOM BIE-2000; Radziejewska et al. 2001b). The two visits resulted in collection of sediment samples used to determine changes in the meiobenthos response variables, except that the number of those variables determined on the 2000 visit was lower, as no taxonomic identification of nematodes or harpacticoids was performed. On the second visit, in April–May 1997, photo- and video-surveys showed a considerable weathering of the Disturber tow tracks (Fig. 4.5; Tkatchenko and Radziejewska 1998). Although still visible, the furrows mentioned above turned out to be considerably eroded, possibly by strong near-bottom currents known to occasionally occur in the area (cf. Kontar and Sokov 1994), and by epifaunal activity, as evidenced by numerous trails of animal movement on the seafloor (Fig. 4.5). In addition, images of the sediment surface showed patches of phytodetritus, its presence being confirmed—as already mentioned—by fluorometrically measuring the sediment phytopigment (chlorophyll *a* and phaeopigments) contents (3.07–4.75 µg/g dry sediment) in two random cores (Radziejewska 2002, Radziejewska et al. 2001b). Interestingly, the pigment pool in one of the samples contained 2.7 % chlorophyll *a*, a labile pigment which is a tracer of fresh, non-degraded algal material present in the sediment (Mincks et al. 2005) suggesting its fast transfer from the water column surface down to the abyss. As already mentioned (cf. Sect. 3.5 and Fig. 3.4), intact diatom cell could still be visible in the sediment samples examined under the microscope. Phytodetritus patches were

Fig. 4.5 IOM BIE test site in 1997: a weathered disturber track (*right-hand* side of the image); note numerous traces of epifaunal activity (*Photo* courtesy of IOM)

particularly well-visible on the sediment surface of cores collected from the formerly impacted area and in the so-called resedimentation zone (R). This station category was introduced to the study following the observation, based on the analysis of contents of sediment traps in moorings deployed around the tow track, that redeposition of sediment suspended as a result of disturbance primarily affected a zone adjacent to the tow track. Thus, samples (collected at randomly selected sampling stations) from that zone, were assigned to this category as, although technically they originated from a non-disturbed area, they did receive some impact from the activity. It should be remembered that the extent of disturbance produced in JET was assessed from such resedimentation zones (cf. Sect. 4.3.5).

Overall, the IOM BIE produced 4 series of data (cf. Table 4.1). The first series (1995) consisted of data pertaining to the pre-impact benthic environment, including the meiobenthos metrics, from 6 cores (collectively termed C1). The second, immediate post-impact series (1995) contained data from inside of the tow track (the disturbance zone, termed Pi1) analysed on 11 cores. The third and the fourth series, obtained 22 months and 5 years after the disturbance (1997 and 2000, respectively), consisted of samples from the unimpacted area of the test site (C2 and C3, 3 cores each), from the tow track zone (Pi2 and Pi3, 5 and 3 cores, respectively), and from the resedimentation area (R1 and R2, 3 cores each). It has to be mentioned that all the sampling stations were selected at random, independently during each cruise, within the prescribed areas (control, impacted, and resedimentation zone).

As shown by Radziejewska (2002), the abundance of meiobenthic communities sampled immediately after the disturbance in 1995 (Pi1: 110.46 ± 60.93 inds/10 cm^2) was, on the average, about half of that sampled prior to the disturbance (C1: 201.60 ± 124.52 inds/10 cm^2), the difference being, however, not significant owing to the high variance in the C1 data. Therefore, the impact on meiobenthos abundance—although discernible—could not be regarded as statistically confirmed. However, examination of the vertical distribution of the Pi1 cores showed the meiobenthos in two of them to differ considerably from the overall pattern found both in the control and impacted area samples (a decrease in the meiobenthos abundance with depth, particularly sharp below 3 cm in the sediment; Radziejewska 2002). One of the two corers (cf. Fig. 2.4a in Radziejewska 2002) supported a severely diminished meiobenthic assemblage, while the distribution of the meiobenthos in the other (cf. Fig. 2.4b in Radziejewska 2002) mirrored the sediment overturning, the density of meiobenthic organisms distinctly increasing downcore, as if the least abundant meiobenthos from the bottom part of the core was brought up to the surface and exposed. Taken together, the results of the first two series of data suggest that the impact of the Disturber, while severe in some places of the tow track zone, left "islands" of unaltered sediment in which meiobenthic assemblages were not affected. Incidentally, a somewhat similar suggestion was put forth by Trueblood et al. (1997) who, however, referred to larger-scale patches of undisturbed seafloor that are created by unmineable portions of the mine sites. Similarly, Gallucci et al. (2008), applying the concept of the patch-mosaic model, referred to patches of undisturbed sediment still remaining in an otherwise disturbed area and retaining subsequent meiobenthic colonizers of the altered sediment.

In the follow-up cruise to the IOM BIE site in April–May 1997, the meiobenthos abundances were observed to increase in both the control (192.13 $\pm$ 36.32 inds/10 cm^2) and the formerly impacted areas (184.83 $\pm$ 85.42 inds/10 cm^2), compared to the respective categories during the 1995 cruise. Particularly high was the mean abundance found in the resedimentation zone (288.21 $\pm$ 65.95 inds/10 cm^2), significantly different from the abundances found in the remaining treatment categories (Radziejewska 2002). The composition of meiobenthic communities, in terms of relative abundances of the two dominant taxa—nematodes and harpacticoids, did not change in any appreciable way, the relative abundances of both taxa remaining in all station categories within ranges typical of the respective taxon, 80.96–87.99 and 5.53–7.69 % for nematodes and harpacticoids, respectively. However, when examined with respect to the genus richness, the composition of the meiobenthos changed markedly. Overall, the numbers of nematode and harpacticoid genera were lower than the respective numbers found in 1995 (175 vs. 226 genera of nematodes and 43 vs. 45 genera of harpacticoids recorded in 1997 and 1995, respectively); however, at least a part of this genus richness reduction can be explained by a lower sampling effort (11 cores) in 1997 compared to 16 cores in 1995. The rarefaction curves for both harpacticoids and nematodes published by Radziejewska et al. (2001a) did not show any asymptotes, so it may be presumed that the number of taxa identified was related to the number of samples collected. On the other hand, the lowered taxon richness could be also related to the changed dominance structure in both the nematode and harpacticoid taxocoenes: in 1997, the nematodes were overwhelmingly dominated by small desmoscolecids (genera *Desmoscolex* and *Pareudesmoscolex*) and the harpacticoids—by unidentified members of the family Argestidae. Most likely, as already discussed earlier (Sect. 3.5), those genera acted as opportunists exploiting arrival to the seafloor of a new abundant food resource in the form of phytodetritus. The phytodetritus deposition was an expression of natural variability of the deep-sea system that occurred during the time elapsing between the cruises, and which was accidentally "stumbled upon", helping to explain at least a part of the variation recorded in the meiobenthic data. Interestingly, Gallucci et al. (2008) provided experimental evidence in support of the suggestion that deep-sea nematodes are capable of detecting and migrating towards patches of phytodetritus on the seafloor.

Thus, the meiobenthos sampled 22 months after the disturbance could be regarded as recovering numerically, the recovery being aided by the naturally occurring phenomena such as phytodetritus supply; the recovery, however, entailed structural changes in the community.

In the second follow-up cruise in June 2000, the Disturber tracks were still visible (Chung et al. 2002). The total meiobenthic abundances showed no significant differences within or between the treatment categories, nor were they significantly different from the levels found in 1997. However, there was an indication of some structural changes in the assemblages to have occurred between 1997 and 2000 in all the treatment categories. For example, the nematode relative abundance and the harpacticoid abundance in C3 were significantly different from C2 and C1: the nematode relative abundance significantly increased and the harpacticoid abundance significantly decreased in C3, compared to the two preceding series. In addition, the domination of desmoscolecids among the nematodes and of argestids among the harpacticoids was no longer observable. No phytodetritus was detected on or in the sediment, either (Radziejewska et al. 2001b).

To find out if more could be learned from the IOM BIE results, data on meiobenthos-related metrics treated as disturbance response variables were, for the purpose of this book, re-analysed using the non-parametric multidimensional scaling (MDS), a multivariate mathematical technique (Clarke and Warwick 2001) which allows to explore similarities and dissimilarities in the structure of individual objects (e.g. samples representing meiobenthic assemblages) defined by several variables (e.g. abundances or relative abundances of individual taxa or the presence/absence of taxa), to infer patterns, and to formulate hypotheses as to those patterns. The analyses were performed using data from all the four series. Preliminary runs of the analyses showed the meiobenthos in one of the cores (MC10 in the first post-impact series of 1995) to be drastically different from the remaining ones, primarily on account of dramatically reduced total density (cf. above; Radziejewska 2002) to render the MDS plot illegible; consequently, this core was excluded from the abundance-based analysis.

Results of the analyses show the structure of meiobenthic communities, viewed in terms of abundance (the left-hand panel of Fig. 4.6), to have been undergoing fairly distinct temporal changes, the temporal signal overriding the signal induced by treatments (control, post-impact, resedimentation). The only exception is to be found in the first year (1995) when the structure of meiobenthic abundance in the post-impact series can be seen to differ somewhat from the control series (the right-hand panel of Fig. 4.6, left-hand side of the plot).

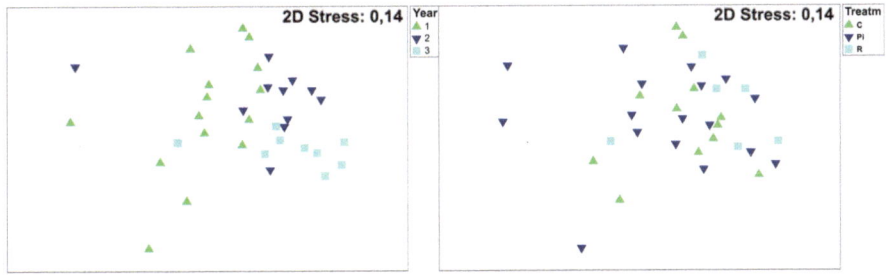

Fig. 4.6 MDS plots showing similarities and dissimilarities between the structure of meioben-thic abundance at individual sampling stations (cores) as analysed in terms of time (years of sampling; *left*) and generalised experimental treatments (C, control; Pi, post-impact; R, resedi-mentation; *right*) in all data series

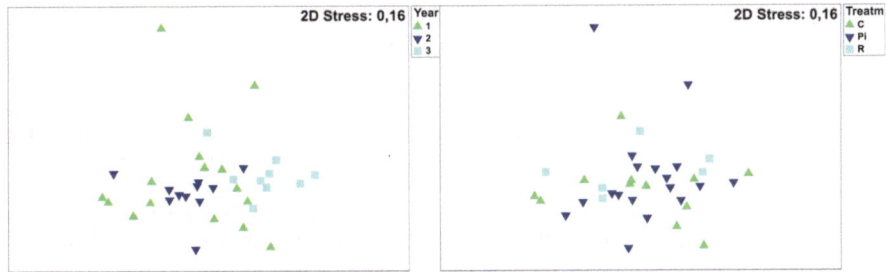

Fig. 4.7 MDS plots showing similarities and dissimilarities between the structure of meiob-enthos relative abundance at individual sampling stations (cores) as analysed in terms of time (years of sampling; *left*) and generalised experimental treatments (C, control; Pi, post-impact; R, resedimentation; *right*) in all data series

Analysis of the meiobenthic community structure viewed in terms of rela-tive abundances of individual taxa (Fig. 4.7) presented a different picture: here, the temporal signal is visible in the separation between years 1 and 2 (1995 and 1997) on the one hand and year 3 (2000) on the other (the left-hand panel of Fig. 4.7). Experimental treatments (the right-hand panel of Fig. 4.7) appear not to have changed the overall meiobenthic community structure, except for two heav-ily impacted stations in year 1 (1995) visible as distinct outliers in that plot of Fig. 4.7. The outliers represent meiobenthos collected from inside the tow track, at a station with a drastically reduced abundance, excluded from the abundance-based analysis above, and at a station with the overturned sediment.

As shown by the (dis)similarity analysis of the nematode and harpacticoid taxo-nomic structure (Fig. 4.8), the two taxa responded differently to the forcing fac-tors presumed to act during the period of study, the experimental treatments and the natural temporal changes of the environment. The taxonomic structure of the nematodes showed a distinct temporal signal (the left-hand panel, upper row of Fig. 4.8) visible as a pronounced separation of points representing the 2 years

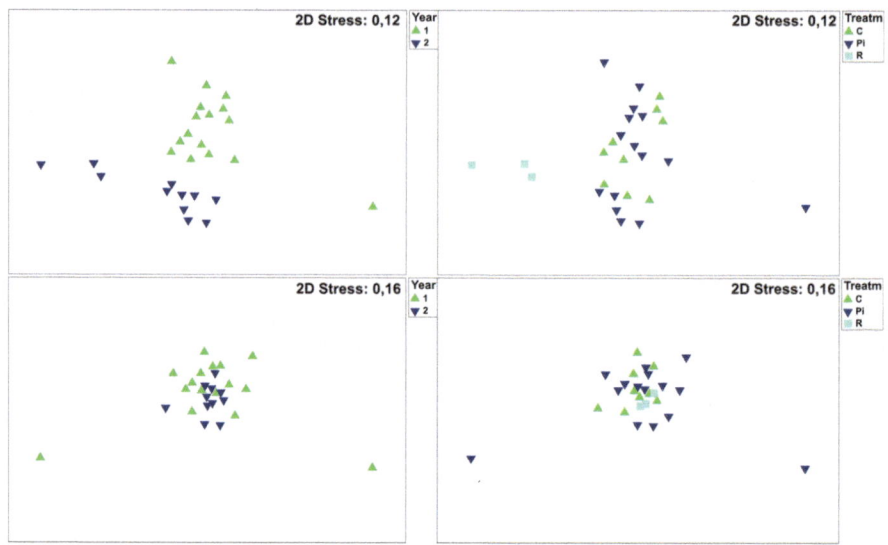

Fig. 4.8 MDS plots showing similarities and dissimilarities between the structure of nematode (*upper panels*) and harpacticoid copepod (*bottom panels*) taxocoenes occurring in the uppermost 3 cm sediment layer (presence/absence of genera) at individual sampling stations (cores) as analysed in terms of time (years of sampling; *left-hand panel*) and generalised experimental treatments (C, control; Pi, post-impact; R, resedimentation; *right-hand panels*) in data series of 1995 and 1997

of study. Although, however, the temporal signal was of the utmost importance, the treatment effect on the nematode taxonomic structure could be seen as well (the right-hand panel, upper row of Fig. 4.8)—in both years the control and post-impact data points are fairly clearly separated. The most visibly separated are, however, the resedimentation zone assemblages, with the lowest genus richness and the most pronounced dominance structure (Radziejewska et al. 2001a). The taxa most responsible for the separation here included the desmoscolecid genera *Pareudesmoscolex* and *Desmoscolex*. It remains unclear, however, how much of that separation depended on differences in the sampling effort (number of cores collected).

In contrast, no clear temporal or treatment-induced effect could be seen in the harpacticoid taxocoene (bottom panels in Fig. 4.8), except for the two 1995 post-impact cores mentioned above, which produced the outliers visible in both plots.

The multivariate analyses presented above seem to reinforce the inferences drawn from the analysis of the univariate data (Table 4.1), particularly with respect to temporal changes in the meiobenthic community structure, but also regarding effects of disturbance. The impact produced by the Disturber was of a moderate magnitude, particularly compared to the impact created in previous BIEs. This is most probably why the effects of the impact in IOM BIE were evident primarily in some most severely altered localities in the tow track, and were less discernible (at least as far as the statistical significance is concerned) in the whole data sets.

Table 4.1 Comparison of results produced by univariate and multivariate analyses of meiobenthic data from IOM BIE

Variable or parameter determined for object (sample)	Univariate analysis (Radziejewska 2002; Radziejewska et al. 2001b)	Multivariate analysis (assemblage structure; this work)	Remarks
Meiobenthos abundance (4 series of data)	Visible, but not significant difference between C1 and Pi1; significant difference between R1 and R2	Temporal signal distinct (difference between years); treatment effect visible in 1995 series	
Meiobenthos relative abundance structure (4 series of data)	Percentage contribution of nematodes significantly different between all control data sets (C1, C2, C3); % contribution of harpacticoids: significantly different between R1 and R2	Temporal signal visible between 1995 + 1997 and 2000; treatment effect visible at two 1995 stations only	
Nematode abundance (4 series of data)	Visible, but not significant difference between C1 and Pi1; significantly different between R1 and R2		
Nematode taxonomic structure 1995 and 1997 data)		Temporal signal distinct; treatment effect visible in 1997 series	Difference, at least partly, attributed to reduced sampling effort in 1997 and 2000, compared to 1995
Harpacticoid abundance (4 series of data)	Significantly different between all control data sets (C1, C2, C3) and between all post-impact data sets (Pi1, Pi2, Pi3)		
Harpacticoid taxonomic structure (1995 and 1997)		Temporal signal indistinct; treatment effect visible in 1995 series	Strong domination of unidentified harpacticoids

Nevertheless, those effects were visible in the meiobenthos, and are highlighted by the results of multivariate analyses (total meiobenthos abundance, nematode community structure). The effects, however, seemed to be of a short duration, particularly with respect to faunal densities, as 22 months post-disturbance the density reduction was no longer visible. In addition, the IOM BIE data highlight changes in the meiobenthic community structure taking place regardless of possible disturbance effects, and likely to have been initially triggered by natural phenomena. Therefore, caution would be in order when interpreting results of sampling campaigns carried out at long time intervals if there is no information how the sedimentary and water column environments were changing in between.

References

Adams WJ, Kimerle RA, Barnett JW (1992) Sediment quality and aquatic life assessment. Environ Sci Technol 26:1864–1875

Ahnert A, Schriever G (2001) Response of abyssal copepoda harpacticoida (crustacea) and other meiobenthos to an artificial disturbance and its bearing on future mining for polymetallic nodules. Deep Sea Res II 48:779–3794

Aller JY (1997) Benthic community response to temporal and spatial gradients in physical disturbance within a deep-sea western boundary region. Deep-Sea Res I 44:39–69

Alongi DM (1985) Effect of physical disturbance on population dynamics and trophic interactions among microbes and meiofauna. J Mar Res 43:351–364

Anonymus (1987) Chapter 6 Environmental considerations. In: Marine minerals: exploring our new ocean frontier. US Congress, Office of Technology Assessment, OTA-0-342, Washington, D.C., 215-245 (accessed via www.vws.princeton.edu)

Austen MC, McEvoy AJ (1997) The use of offshore meiobenthic communities in laboratory microcosm experiments: response to heavy metal contamination. J Exp Mar Biol Ecol 211:247–261

Austen MC, Widdicombe S, Villano-Pitacco N (1998) Effects of biological disturbance on diversity and structure of meiobenthic nematode communities. Mar Ecol Prog Ser 174:33–246

Baker CM, Bett BJ, Billett DSM et al (2001) An environmental perspective. In: WWF/IUCN, the status of natural resources on the high seas. WWF/IUCN, Gland, Switzerland

Balsamo M, Semprucci F, Frontalini F et al (2012) Meiofauna as a tool for marine ecosystem biomonitoring. In: Cruzado A (ed) Marine ecosystems. InTech, doi: 10.5772/34423

Barnes DKA (1999) The influence of ice on polar nearshore benthos. J Mar Biol Ass UK 79:401–407

Barnes B, Sidhu HS, Roxburgh H (2006) A model integrating patch dynamics, competing species and the intermediate disturbance hypothesis. Ecol Model 194:414–420

Barnett B, Yamauchi H (1995) Deep sea sediment resuspension system used for the Japan deep sea impact experiment. In: Yamazaki T, Aso K, Okano Y et al. (eds) Proceedings of ISOPE—Ocean Mining Symposium, pp 175–179, Tsukuba, Japan, 21–22 Nov 1995

Berge S, Markussen JM, Vigerust G (1991) Environmental consequences of deep seabed mining. Problem areas and regulations. Fridtjof Nansen Institute, Lysaker

Bett BJ, Narayanaswamy BE (2014) Genera as proxies for species α- and β-diversity: tested across a deep-water Atlantic-Arctic boundary. Mar Ecol doi: 10.1111/maec.12100

Boesch DF, Rosenberg R (1981) Response to stress in marine benthic communities. In: Barrett GW, Rosenberg R (eds) Stress effects on natural ecosystems. Wiley, New York

Borja A, Muxika I (2005) Guidelines for the use of AMBI (AZTI's Marine Biotic Index) in the assessment of the benthic ecological quality. Mar Poll Bull 50:787–789

Borowski C (2001) Physically disturbed deep-sea macrofauna in the Peru Basin, southeast Pacific, revisited 7 years after the experimental impact. Deep-Sea Res II 48:3809–3839

Brockett T, Richards CZ (1994) Deepsea mining simulator for environmental impact studies. Sea Technol 35(8):77–82

Burd BJ (2002) Evaluation of mine tailings effects on a benthic marine infaunal community over 29 years. Mar Environ Res 53:481–519

Cadotte MW (2007) Competition-colonization trade-offs and disturbance effects at multiple scales. Ecology 88:823–829

Cardinale BJ, Nelson K, Palmer MA (2000) Linking species diversity to the functioning of ecosystems: on the importance of environmental context. Oikos 91:175–183

Carman KR, Thistle D, Fleeger JW et al (2004) Influence of introduced CO_2 on deep-sea metazoan meiofauna. J Oceanog 60:767–772

Chung JS (2013) Commercial mining system development for manganese nodules: take direct-to- or incremental-to-5,000-m approach? In: Chung JS, Komai T (eds) Proceedings of 10th ISOPE Ocean Mining Gas Hydrates Symposium, pp 1–4, Szczecin, Poland, 22–26 Sep 2013

Chung JS, Schriever G, Sharma R et al (eds) (2002) Deep seabed mining environment: preliminary engineering and environmental assessment. ISOPE Spec Rep OMS-EN-1, ISOPE, Cupertino, California, USA

Clarke KR, Warwick RM (2001) Change in marine communities: an approach to statistical analysis and interpretation, 2nd edn. Primer-E, Plymouth

Colangelo MA, Macrí T, Ceccherelli VU (1996) A field experiment on the effect of two types of sediment disturbance on the rate of recovery of a meiobenthic community in a eutrophicated lagoon. Hydrobiologia 329:57–67

Connell JH (1978) Diversity in tropical rain forests and coral reefs. Sci New Ser 199:1302–1310

Cowie PR, Widdicombe S, Austen MC (2000) Effects of physical disturbance on an estuarine intertidal community: field and mesocosm results compared. Mar Biol 136:485–495

Creed EL, Coull BC (1984) Sand dollar, *Mellita quinquiesperforata* (Leske), and sea pansy, *Renilla reniformis* (Cuvier): effects on meiofaunal abundance. J Exp Mar Biol Ecol 84:225–234

Danovaro R (2000) Benthic microbial loop and meiofaunal response to oil-induced disturbance in coastal sediments: a review. Inter J Environ Poll 13:380–391

Dye AH, Lasiak TA (1986) Microbenthos, meiobenthos and fiddler crabs: trophic interactions in a tropical mangrove sediment. Mar Ecol Prog Ser 32:259–264

Fleeger JW, Shirley TC, Carls MG et al (1996) Meiofaunal recolonization experiment with oiled sediments. Am Fish Soc Symp 18:271–285

Flentje W, Lee SE, Virnovskaia A et al (2012) Polymetallic nodule mining: innovative concepts for commercialisation. University Southampton, LRET Collegium 2012 Series, 5

Fontaubert AC de (2001) Legal and political considerations. In: WWF/IUCN, the status of natural resources on the high-seas. WWF/IUCN, Gland, Switzerland

Franz DR, Friedman I (2002) Effects of a macroalgal mat (*Ulva lactuca*) on estuarine sand flat copepods: an experimental study. J Exp Mar Biol Ecol 271:209–226

Fukushima T (1995) Overview "Japan Deep-Sea Impact Experiment = JET". In: Yamazaki T, Aso K, Okano Y et al (eds) Proceedings of ISOPE—Ocean Mining Symposium, pp 47–53, Tsukuba, Japan, 21-22 Nov 1995

Gallucci F, Moens T, Vanreusel A et al (2008) Active colonisation of disturbed sediments by deep-sea nematodes: evidence for the patch mosaic model. Mar Ecol Prog Ser 367:173–183

Glasby GP (2000) Lessons learned from deep-sea mining. Science 289:551–553

Glasby GP (2002) Deep seabed mining: past failures and future prospects. Mar Geores Geotechnol 20:161–176

Glover AG, Smith CR (2003) The deep-sea floor ecosystem: current status and prospects of anthropogenic change by the year 2025. Environ Conserv 30:219–241

Grassle JF, Morse-Porteous LS (1987) Macrofaunal colonization of disturbed deep-sea environments and the structure of deep-sea benthic communities. Deep-Sea Res 34:1911–1950

Gray JS, Elliott M (2009) Ecology of marine sediments. From science to management, 2nd edn. Oxford University Press, Oxford

Green RH (1979) Sampling design and statistical methods for environmental biologists. Wiley, New York

Halpern BS, Selkoe KA, Michell F et al (2007) Evaluating and ranking the vulnerability of global marine ecosystems to anthropogenic threats. Conserv Biol 21:1301–1315

Hein JR, Petersen S (2013) The geology of manganese nodules. In: Baker E, Beaudoin Y (eds) Deep sea minerals: manganese nodules, a physical, biological, environmental, and technical review, vol 1B. Secretariat of the Pacific Community, GRID-Arendal

Hill RA, Chapman PM, Mann GS et al (2000) Level of detail in ecological risk assessments. Mar Poll Bull 40:471–477

Hobbs CH (2002) An investigation of potential consequences of marine mining in shallow water: an example from the mid-Atlantic coast of the United States. J Coast Res 18:94–101

Huxham M, Roberts I, Bremner J (2000) A field test of the intermediate disturbance hypothesis in the soft-bottom intertidal. Int Rev Hydrobiol 85:379–394

Jewett SC, Feder HM, Blanchard A (1999) Assessment of the benthic environment following offshore placer gold mining in the northeastern Bering Sea. Mar Environ Res 48:91–122

Jumars PA (1981) Limits to predicting and detecting benthic community responses to manganese nodule mining. Mar Mining 3:213–229

Kaneko (Sato) T, Ogura K, Fukushima T (1995) Preliminary results of meiofauna and bacteria abundance in an environmental impact experiment. In: Yamazaki T, Aso K, Okano Y et al (eds) Proceedings of ISOPE—Ocean Mining Symposium, Tsukuba, Japan, pp 181–186

Kaneko (Sato) T, Maejima Y, Teishima Y (1997) The abundance and vertical distribution of abyssal benthic fauna in the Japan Deep-sea impact experiment. In: Chung JS, Das BM, Matsui T et al (eds) Proceedings of 7th ISOPE Conference, vol 1. Honolulu, USA, pp 475–480

Khripounoff A, Caprais J-C, Crassous P (2006) Geochemical and biological recovery of the disturbed seafloor in polymetallic nodule fields of the Clipperton-Clarion Fracture Zone (CCFZ) at 5, 000-m depth. Limnol Oceanog 51:2033–2041

Kontar EA, Sokov AV (1994) A benthic storm in northeastern tropical Pacific over the fields of manganese nodules. Deep-Sea Res 41:1069–1089

Kotliński R, Stoyanova V (1998) Physical, chemical, and geological changes of marine environment caused by the benthic impact experiment at the IOM BIE site. In: Chung JS, Olagnon M, Kim CH et al (eds) Proceedings of 8th ISOPE Conference, vol 2. Montreal, Canada, pp 277–281

Lambshead PJD, Hodda M (1994) The impact of disturbance on measurements of variability in marine nematode populations. Vie Milieu 44:21–27

Lavering IH (1994) Marine environments of Southeast Australia (gippsland shelf and bass strait) and the impact of offshore petroleum exploration and production activity. Mar Geores Geotechnol 12:201–226

La Rosa T, Mirto S, Mazzola A et al (2001) Differential responses of benthic microbes and meiofauna to fish-farm disturbance in coastal sediments. Environ Poll 112:427–434

Lee HJ, Vanhove S, Peck LS et al (2001) Recolonsation of meiofauna after catastrophic iceberg scouring in shallow Antarctic sediments. Polar Biol 24:918–925

Lodge M, Johnson D, Le Gurun G et al (2014) Seabed mining: international seabed authority environmental management plan for the Clarion-Clipperton Zone. A partnership approach. Mar Pol 49:66–72

Magni P, Hyland I, Manzella G et al (eds) (2005) Proceedings of the workshop "indicators of stress in the marine benthos", Torregrande-Oristano, Italy, Paris, UNESCO/IOC, IMC, 2005. (IOC Workshop Rep 195) (IMC Spec Publ ISBN 88-85983-01-4), 8–9 Oct 2004

Mahatma R (2009) Meiofauna communities of the Pacific nodule province: abundance, diversity and community structure. Ph.D. Thesis, University of Oldenburg, Oldenburg, Germany

Markussen JM (1994) Deep seabed mining and the environment: consequences, perceptions, and regulations. In: Bergesen HO, Parmann G (eds) Green globe yearbook of international co-operation on environment and development 1994. Oxford University Press, Oxford

Martin J, Sanchez-Cabeza JA, Eriksson M et al (2009) Recent accumulation of trace metals in sediments at the DYFAMED site (Northwestern Mediterranean Sea). Mar Poll Bull 59:146–153

Miljutin DM, Miljutina MA, Martinez Arbizu P et al (2011) Deep-sea nematode assemblage has not recovered 26 years after experimental mining of polymetallic nodules (Clarion-Clipperton Fracture Zone, Tropical Eastern Pacific). Deep Sea Res I 58:885–897

Mincks SL, Smith CR, DeMaster DJ (2005) Persistence of labile organic matter and microbial biomass in Antarctic shelf sediments: evidence of a sediment food bank. Mar Ecol Prog Ser 300:3–19

Mirto S, La Rosa T, Gambi C et al (2002) Nematode community response to fish-farm impact in the western Mediterranean. Environ Poll 116:203–214

Morgan CL, Odunton NA, Jones AF (1999) Synthesis of environmental impacts of deep seabed mining. Mar Geores Geotechnol 17:307–356

Mullineaux LS (1987) Organisms living on manganese nodules and crusts: distribution and abundance at three North Pacific sites. Deep-Sea Res 34:165–184

Mullineaux LS (1988) The role of settlement in structuring a hard-substratum community in the deep sea. J Exp Mar Biol Ecol 120:247–261

National Research Council (1984) Deep seabed stable reference areas. National Academic Press, Washington, D.C

Nilsson C, Grelsson G (1995) The fragility of ecosystems: a review. J Appl Ecol 32:677–692

Ozturgut E, Anderson GD, Burns RE et al (1978) Deep ocean mining of manganese nodules in the North Pacific: pre-mining environmental conditions and anticipated mining effects. NOAA Techn Mem, ERL MESA-33

Ozturgut E, Lavelle JW, Burns RE (1981) Impacts of manganese nodule mining on the environment: results from pilot-scale mining tests in the North equatorial Pacific. In: Geyer RA (ed), Marine environmental pollution. 2. Dumping and Mining. Elsevier, Oceanography Series 27B

Pearson TH (1981) Stress and catastrophe in marine benthic ecosystems. In: Barrett GW, Rosenberg R (eds) Stress effects on natural ecosystems. Wiley, New York

Pearson TH, Rosenberg R (1978) Macrobenthic succession in relation to organic enrichment and pollution of the marine environment. Oceanog Mar Biol Ann Rev 16:229–311

Pogrebov VB, Fokin SI, Galtsova VV et al (1997) Benthic communities as influenced by nuclear testing and radioactive waste disposal off Novaya Zemlya in the Russian Arctic. Mar Poll Bull 35:333–339

Powell EN, Bright TJ, Woods A et al (1983) Meiofauna and the thiobios in the east flower garden brine seep. Mar Biol 72:269–283

Radziejewska T (2002) Responses of deep-sea meiobenthic communities to sediment disturbance simulating effects of polymetallic nodule mining. Int Rev Hydrobiol 87:459–479

Radziejewska T, Masłowski J (1997) Macro- and meiobenthos of the Arkona Basin (western Baltic Sea): differential recovery following hypoxic events. In: Hawkins LE, Hutchinson S, Jensen AC et al (eds) The responses of marine organisms to their environments. Proceedings 30th European marine biology symposium, University of Southampton, Southampton

Radziejewska T, Drzycimski I, Galtsova VV et al (2001a) Changes in genus-level diversity of meiobenthic free-living nematodes (Nematoda) and harpacticoids (Copepoda Harpacticoida) at an abyssal site following experimental sediment disturbance. In: Chung JS, Stoyanova V (eds) Proceedings of 4th Ocean Mining Symposium, Szczecin, Poland, pp 38–43

Radziejewska T, Rokicka-Praxmajer J, Stoyanova V (2001b) IOM BIE revisited: meiobenthos at the IOM BIE site 5 years after the experimental disturbance. In: Chung JS, Stoyanova V (eds) Proceedings of 4th Ocean Mining Symposium, Szczecin, Poland, pp 63–68

Ramirez-Llodra E, Tyler PA, Baker MC et al (2011) Man and the last great wilderness: human impact on the deep sea. PLoS ONE 6:e22588

Rees HL, Boyd SE, Schratzberger M et al (2006) Role of benthic indicators in regulating human activities at sea. Environ Sci Policy 9:496–508

Rhoads DC (1974) Organism-sediment relations on the muddy sea floor. Oceanog Mar Biol Ann Rev 12:263–300

Roxburgh SH, Shea K, Wilson JB (2004) The intermediate disturbance hypothesis: patch dynamics and mechanisms of species coexistence. Ecology 85:359–371

Savage C, Field JG, Warwick RM (2001) Comparative meta-analysis of the impact of offshore marine mining on macrobenthic communities versus organic pollution studies. Mar Ecol Prog Ser 221:265–275

Schratzberger M, Warwick RM (1998) Effects of physical disturbance on nematode communities in sand and mud: a microcosm experiment. Mar Biol 130:643–650

Schratzberger M, Dinmore TA, Jennings S (2002) Impacts of trawling on the diversity, biomass and structure of meiofauna assemblages. Mar Biol 14:83–93

Schratzberger M, Rees HL, Boyd SE (2000) Effects of simulated deposition of dredged material on structure of nematode assemblages–the role of contamination. Mar Biol 137:613–622

Schriever G (1995) DISCOL—disturbance and recolonization experiment of a manganese nodule area of the Southeastern Pacific. In: Yamazaki T, Aso K, Okano Y et al (eds) Proceedings of ISOPE—ocean mining symposium, pp 163–166, Tsukuba, Japan, 21–22 Nov 1995

Schriever G, Bussau C, Thiel H (1991) DISCOL–precautionary environmental impact studies for future manganese nodule mining and first results on meiofauna abundance. Proc Adv Mar Technol Conf 4:47–57

Schriever G, Ahner A, Bluhm H et al (1997) Results of the large-scale deep-sea experimental study DISCOL during eight years of investigation. In: Chung JS, Das BM, Matsui T, Thiel H (eds) Proceedings of 7th ISOPE Conference, vol 2. Honolulu, Hawaii, pp 438–444

Sherman KM, Coull BC (1980) The response of meiofauna to sediment disturbance. J Exp Mar Biol Ecol 46:59–71

Sherman KM, Reidenauer JA, Thistle D et al (1983) Role of a natural disturbance in an assemblage of marine free-living nematodes. Mar Ecol Prog Ser 11:23–30

Shirayama Y (1999) Biological results of the JET project: an overview. In: Chung JS, Sharma R (eds) Proceedings of 3rd Ocean Mining Symposium, pp 185–190, Goa, India, 8–10 Nov 1999

Smith CR, Levin LA, Koslow A et al (2008) The near future of the deep seafloor ecosystems. In: Polunin N (ed) Aquatic ecosystems: trends and global prospects. Cambridge Univ Press, Cambridge

Smith EP (2002) BACI design. In: El-Shaarawi A, Piegorsch WW (eds) Encyclopedia of environmetrics, vol 1. Wiley, Chichester

Smith S, Heydon R (2013) Processes related to the technical development of marine mining. In: Baker E, Beaudoin Y (eds) Deep sea minerals. Manganese nodules, a physical, biological, environmental and technical review, vol 1B. Secretariat of the Pacific Community, GRID-Arendal

Somerfield PJ, Rees HL, Warwick RM (1995) Interrelationships in community structure between shallow-water marine meiofauna and macrofauna in relation to dredgings disposal. Mar Ecol Prog Ser 127:13–112

Taguchi K, Nakata K, Aoki S et al (1995) Environmental study on the deep-sea mining of manganese nodules in the northeastern tropical Pacific. In: Yamazaki T, Aso K, Okano Y et al (eds) Proceedings of ISOPE—Ocean Mining Symposium, pp 167–174, Tsukuba, Japan, 21–22 Nov 1995

Thiel H (2001) Use and protection of the deep sea–an introduction. Deep-Sea Res II 48:3427–3431

Thiel H, Foell EJ, Schriever G (1992) Potential environmental effects of deep seabed mining. Univ Hamburg, Hamburg, 26

Thiel H, Forschungsverbund Tiefsee-Umweltschutz (1995) The german environmental impact research for manganese nodule mining in the SE Pacific Ocean. In: Yamazaki T, Aso K, Okano Y et al (eds) Proceedings of ISOPE—Ocean Mining Symposium, pp 39–45, Tsukuba, Japan, 21–22 Nov 1995

Thiel H, Forschungsverbund Tiefsee-Umweltschutz (2001) Evaluation of the environmental consequences of polymetallic nodule mining based on the results of the TUSCH Research Association. Deep-Sea Res II 48:3433–3452

Thiel H, Angel MV, Foell EJ et al (1998) Environmental risks from large-scale ecological research in the deep-sea; a desk study. Office for Official Publications of the European Communities, Luxembourg

Thistle D (1998) Harpacticoid copepod diversity at two physically reworked sites in the deep sea. Deep-Sea Res II 45:13–24

Thrush S, Dayton PK (2002) Disturbance to marine benthic habitats by trawling and dredging: implications for marine biodiversity. Ann Rev Ecology System 33:449–473

Tkatchenko GG, Radziejewska T (1998) Recovery and recolonization processes in the area disturbed by a polymetallic nodule collector simulator. In: Chung JS, Olagnon M, Kim CH et al (eds) Proceedings of 8th ISOPE Conference, vol 2. Montreal, Canada, pp 282–286

Trueblood DD, Ozturgut E (1997) The benthic impact experiment: a study of the ecological impacts of deep seabed mining on abyssal benthic communities. In: Chung JS, Das BM, Matsui T, Thiel H (eds) Proceedings of 7th (1997) ISOPE Conference, pp 481–487, Honolulu, USA, 25–30 May 1997

Trueblood DD, Ozturgut E, Pilipchuk M et al (1997) The ecological impact of the joint U.S.–Russian benthic impact experiment. In: Proceedings of 2nd Ocean Mining Symposium, pp 139–145, Seoul, Korea, 24-26 Nov 1997

Underwood AJ (1992) Beyond BACI: the detection of environmental impacts on populations in the real, but variable, world. J Exp Mar Biol Ecol 161:145–178

Underwood AJ (1996) Detection, interpretation, prediction and management of environmental disturbances: some roles for experimental marine ecology. J Exp Mar Biol Ecol 200:1–27

Varon R, Thistle D (1988) Response of a harpacticoid copepod to a small-scale natural disturbance. J Exp Mar Biol Ecol 118:245–256

Veillette J, Juniper SK, Gooday AJ et al (2007a) Influence of surface texture and microhabitat heterogeneity in structuring nodule faunal communities. Deep-Sea Res I 54:1936–1943

Veillette J, Sarrazin J, Gooday AJ et al (2007b) Ferromanganese nodule fauna in the Tropical North Pacific Ocean: species richness, faunal cover and spatial distribution. Deep-Sea Res I 54:1912–1935

Vopel K, Thiel H (2001) Abyssal nematode assemblages in physically disturbed and adjacent sites of the eastern equatorial Pacific. Deep-Sea Res II 48:3795–3808

Warwick RM (1993) Environmental impact studies on marine communities: Pragmatical considerations. Austral J Ecol 18:63–80

Warwick RM, Clarke KR (1993a) Comparing the severity of disturbance: a meta-analysis of marine macrobenthic community data. Mar Ecol Prog Ser 92:221–231

Warwick RM, Clarke KR (1993b) Increased variability as a symptom of stress in marine communities. J Exp Mar Biol Ecol 172:215–226

Warwick RM, Clarke KR (1994) Relearning the ABC: taxonomic changes and abundance/biomass relationships in disturbed benthic communities. Mar Biol 118:739–744

Warwick RM, Clarke KR (1995) New 'biodiversity' measures reveal a decrease in taxonomic distinctness with increasing stress. Mar Ecol Prog Ser 129:301–305

Warwick RM, Clarke KR (2001) Practical measures of marine biodiversity based on relatedness of species. Oceanog Mar Biol Ann Rev 39:207–231

Warwick RM, Clarke KR, Gee JM (1990) The effect of disturbance by soldier crabs *Mictyris platycheles* H. Milne Edwards on meiobenthic community structure. J Exp Mar Biol Ecol 135:19–33

Widdicombe S, Austen MC (2001) The interaction between physical disturbance and organic enrichment: An important element in structuring benthic communities. Limnol Oceanog 46:1720–1733

Wilson GDF (1987) Crustacean communities of the manganese nodule province (DOMES site A compared with DOMES site C). Report for the National Oceanic and Atmospheric Administration Office of Ocean and Coastal Resource Management (Oceans and Energy) on Contract NA-84-ABH-0030 (accessed via www.personal.usyd.edu.au/~buz/PDF/Crustacean_Communities1999ver.pdf)

Yamazaki T, Kajitani J (1999) Deep-sea environment and impact experiment to It. In: Chung JS, Matsui T, Koterayama W (eds) Proceedings of 9th ISOPE Conference, vol 1. Brest, France, pp 374–381

Yamazaki T, Sharma R (2001) Estimation of sediment properties during benthic impact experiments. Mar Geores Geotechnol 19:269–299

Chapter 5
Epilogue

At the time of this writing (2014), it has been almost 20 years since the last of the series of experiments in the nodule fields (BIEs) was conducted. As the time went by, the interest in nodule development waxed and waned, and other deep-sea metal-liferous resource types were intermittently attracting attention of the mining and scientific communities. The discovery of abundant metalliferous muds in the Pacific (Kato et al. 2011), a promise of an easier access to cobalt-rich crusts (Schlacher et al. 2013), and the metal resources to be tapped from hydrothermal vent smokers (e.g. massive sulphides; Boschen et al. 2011; Gena 2013; Van Dover 2011), to mention just a few, periodically shifted the limelight of interest from the nodules as the potentially most valuable deep-sea metal repository to other metal "treasure troves". Invariably, however, the interest returned to the nodules, which—or the promise they offered—even made their way to the popular fiction (Bagley 1984).

The concerns raised by future nodule exploitation (Glover and Smith 2003) and addressed by the impact experiments have prompted a number of authors and research groups to undertake reviews of outcomes of these experiments. While some, like Oebius et al. (2001), and Yamazaki and Sharma (2001) focused on alterations in the sedimentary milieu as elucidated by the experiments, and others (e.g. Chung et al. 2002; Sharma 2005) drew preliminary conclusions with respect to engineering aspects, a comprehensive ecosystem perspective was adopted by Morgan et al. (1999) and Thiel (2001). The two latter teams, however, realised the limitations of knowledge on deep seafloor biota necessary to adequately address the environmental impact of mining. Morgan et al. (1999) provided a fairly detailed account of the oceanographic conditions in nodule fields, including those in CCFZ, but were not able to come up with equally detailed treatment of biological variables, particularly the benthic communities. They wrote that "Adequate characterization of the benthic communities in the Clarion-Clipperton region [...] has not yet been achieved" (p. 329).

As could be seen from the preceding parts of this book, important steps towards such characterization have been taken in association with impact experiments, and our knowledge has grown since the time Morgan et al. (1999) wrote their synthesis, as later work did provide a substantial body of data on the benthic biota.

© The Author(s) 2014
T. Radziejewska, *Meiobenthos in the Sub-equatorial Pacific Abyss*,
SpringerBriefs in Earth System Sciences, DOI 10.1007/978-3-642-41458-9_5

This, however, does not mean that the knowledge is advanced enough to satisfy the needs of informed management of deep seabed mineral resources (Lodge et al. 2014, Mengerink et al. 2014). In their evaluation of ecosystem consequences of nodule mining, Thiel and Forschungsverbund Tiefsee-Unmweltschutz (2001) wrote that, although some impacts (particularly in the uppermost sediment layers and the near-bottom water) were unavoidable, they did not encounter strong arguments against nodule mining, provided the impacts are kept at the minimum level. They did, however, point out that many questions, particularly those pertaining to the re-establishment of geochemically and ecologically stable conditions after commercial mining disturbance remain open.

The concept of ecologically stable conditions entails recolonisation of the disturbed sediment by the communities similar in structure and function to the original ones. Therefore, an important aspect of the impact experiments was the assessment of the recolonisation rates. Thiel and Forschungsverbund Tiefsee-Unmweltschutz (2001), like Grassle (1977) before them, drew attention to the slow recolonisation of deep-sea sediments after defaunation. However, as pointed out by Grassle and Morse-Porteous (1987), recolonisation potential is enhanced or inhibited by various environmental variables and differs strongly among various taxa.

In this context, the deep-sea meiobenthos, a "hidden but significant" (Ramirez-Llodra et al. 2010) component of the sedimentary biota in the abyss, was included in (almost) all the impact experiments performed to date. Based on the experience from shallow areas (Heip 1980), the meiobenthos was considered to represent one of the most useful vectors (if not the most useful one) of biological response variables. In the preceding parts, an attempt was made to extract the relevant information from the existing evidence. Now the time has come to answer the question whether the meiobenthos meets the expectations of being a sufficiently suitable tool in assessment of environmental quality alteration (Thistle 2003).

This tool can be used, as already pointed out, to elucidate two interrelated aspects of impact: severity and persistence. The severity of impact can be inferred from a reduction in the meiobenthos assemblage abundance and change of its composition, particularly with respect of the taxon richness. However, the actual type of impact has to be taken into account, as individual studies that were discussed in this book differed in the impact type they addressed. While DISCOL and IOM BIE, particularly the latter, focused on changes in the seafloor directly physically affected by the disturbance-producing device (e.g. Disturber runner grooves in IOM BIE), Joint US-Russian BIE and JET concentrated on the impact understood as sediment blanketing by resedimentation (burial). In both types of impacts, the meiobenthic communities as a whole and their major components—nematodes, and in certain instances also harpacticoid copepods—did show undisputable local responses visible, immediately post-disturbance, as reduced abundances in the area directly affected by the disturbance-producing device (DISCOL, IOM BIE) and in the area of resedimentation (BIE, JET). On a longer term, some community functions were observed to have been altered (e.g. changed proportions between nematode trophic guilds observed in DISCOL or inferred in JET). However, no general direct measure of impact severity from changed abundances and assemblage composition has been developed so far. As stressed by Morgan et al. (1999), such a measure could be sought in a quantitative relationship between the resedimented layer thickness (burial depth) and faunal change (including succession). No such relationship has been, or could have been, developed in the studies discussed—primarily because the resedimentation thickness could not be actually measured and was only indirectly estimated.

The persistence of impact can be assessed by recording the recolonisation rate. In contrast to the macrobenthos, meiobenthic organisms have been documented, under various conditions including the deep sea, to be fairly fast recolonisers of disturbed sediments. For example, Freese et al. (2012) noted that recolonisation by meiofauna of artificial substrata was completed within 1 year, but found the recolonisation to have been greatly facilitated by abundance of food resources, and to have been accomplished by juvenile (and /or small-sized) meiobenthos. The analogy with the phytodetritus supply and domination of abundant small desmoscolecid nematodes, as observed in IOM BIE, is obvious. Similarly, Gallucci et al. (2008) pointed out to the ability of the meiobenthos, notably the nematodes, to detect—and migrate towards—patches of abundant food.

Thus, the recolonisation of sediments, disturbed by nodule extraction, by the meiobenthos will depend on provision of food resources (as was assumed in JET) and by the presence of a pool of colonisers in the vicinity of the disturbed area. It can be assumed that, in the impact experiments described, the pool of colonisers did exist, also probably within the disturbed area itself ("islands" or patches of unaltered sediment; cf. Gallucci et al. 2008), so the disturbed areas were observed to have been, by and large, recolonised. The actual recolonisation rate, however, could have hardly been evaluated given the long time span between the post-impact and follow-up samplings. Indeed, the intermittency of follow-up cruises seems to have been one of major obstacles in making inferences as to the recolonisation rates displayed by the meiobenthos in the experiments described. The shortest time interval was applied in JET (6 months) and the observations did show a relatively fast rebound of the meiobenthos abundance in the resedimented area. As shown by IOM BIE, even a between-samplings interval of 22 months can result in unexpected effects, such as phytodetritus deposition on the seafloor, intervening (and most likely aiding) the recolonisation. However, while the increased meiobenthic abundance can be ascribed to the fast recolonisation rate, explanation of changes in the community composition, and the associated alteration in functional traits, cannot be made without some knowledge of what actually happened (or could have happened) in-between the sampling campaigns. Therefore, such campaigns, while focusing on the meiobenthos, should include a suite of associated measurements and observations on the abiotic environment (e.g. long-term current data, video footage of megafaunal activity close to and on the seabed, sediment geochemistry with a particular attention to proxies of labile organic matter supply in the form of phytal pigment contents) and relevant biotic components, e.g. microorganisms (bacteria, protists) to, if not record as in the case of currents, try to get a glimpse on past events which would leave traces persisting until the time of sampling.

There are also methodological aspects which should be paid a lot of attention to. The sampling design should allow for the collection of appropriate, not too low, number of samples. Grassle and Morse-Porteous (1987) referred to low-resolution sampling as making it difficult to assess environmental and faunal patchiness in the deep sea. Although it is admittedly difficult to state *a priori* how many samples ought to be collected to obtain a desired level of assessment precision, there are statistical techniques which can aid in determining the number of samples (e.g. Quinn and Keough 2002). These should be coupled with information provided by the existing data.

The most important issue concerns the sediment sampling gear. To adequately deal with tiny organisms such as the meiobenthos, samples of sediment which is as close to being unaltered by coring as possible are most desirable. Therefore the multicorer should be used or, if possible, cores should be collected by a submersible manipulator. Finally, the data analysis should, as is more and more frequently the common practice, combine the univariate and multivariate methods for detecting impact effects and assessing their significance.

Such questions, particularly those pertaining to the logistics of sampling design and a possibly high analytical resolution (i.e. an adequately low level of taxonomic identification—at least to genus; the species level should be applied only if the identifications can be intercalibrated), can be adequately addressed in the international collaboration framework (Lodge et al. 2014). The recommendations of Morgan et al. (1999) concerning unification of mining-related research efforts on benthic communities (common rules for taxonomic reference, a single site of holding of specimen collections, establishment of long-term jointly studied experimental reference sites for intercomparison of research techniques) are still valid, and are particularly pertinent with regard to the meiobenthos.

Based on data accumulated so far and reviewed in this book, it may be contended that the deep-sea meiobenthos, when used in anthropogenic impact evaluation, is indeed a good proxy. Its potential has been, however, under-utilised owing to various constraints stemming from inadequate knowledge as well as from logistic and methodological constraints. The present climate of international scientific collaboration (Lodge et al. 2014) offers an opportunity to overcome those constraints in the interest of good governance over deep-sea areas replete in mineral resources.

References

Bagley D (1984) Night of error. Brockhurst Publications Ltd., London

Boschen RE, Rowden AA, Clark MR et al (2011) Mining of deep-seafloor massive sulfides: a review of the deposits, their benthic communities, impacts from mining, regulatory frameworks and management strategies. Ocean Coast Manag 84:54–67

Chung JS, Schriever G, Sharma R (eds) (2002) Deep seabed mining environment: preliminary engineering and environmental assessment. ISOPE Special Report OMS-EN-1, ISOPE, Cupertino, California, USA

Freese D, Schewe I, Kanzog C et al (2012) Recolonisation of new habitats by meiobenthic organisms in the deep Arctic Ocean: an experimental approach. Polar Biol 35:1801–1813

Gallucci F, Moens T, Vanreusel A et al (2008) Active colonisation of disturbed sediments by deep-sea nematodes: evidence for the patch mosaic model. Mar Ecol Prog Ser 367:173–183

Gena K (2013) Deep sea mining of submarine hydrothermal deposits and its possible environmental impacts in Manus Basin, Papua New Guinea. Proc Earth Planet Sci 6:226–233

Glover AG, Smith CR (2003) The deep-sea floor ecosystem: current status and prospects of anthropogenic change by the year 2025. Environ Conserv 30:219–241

Grassle JF (1977) Slow recolonisation of deep-sea sediment. Nature 265:618–619

Grassle JF, Morse-Porteous LS (1987) Macrofaunal colonization of disturbed deep-sea environments and the structure of deep-sea benthic communities. Deep-Sea Res 34:1911–1950

Heip C (1980) Meiobenthos as a tool in the assessment of marine environmental quality. Rapp Proc-verb Réun 179:182–187

Kato Y, Fujinaga L, Nakamura K et al (2011) Deep-sea mud in the Pacific Ocean as a potential resource for rare-earth elements. Nat Geosci 4:535–539

Lodge M, Johnson D, Le Gurun G et al (2014) Seabed mining: International Seabed Authority environmental management plan for the Clarion–Clipperton Zone. A partnership approach. Mar Policy 49:66–72

Mengerink K, Van Dover CL, Ardron J et al (2014) A call for deep-ocean stewardship. Science 344:696–698

Morgan CL, Odunton NA, Jones AF (1999) Synthesis of environmental impacts of deep seabed mining. Mar Geores Geotechnol 17:307–356

Oebius HU, Becker HJ, Rolinski S et al (2001) Parameterization and evaluation of marine environmental impacts produced by deep-sea manganese nodule mining. Deep-Sea Res II 48:3453–3467

Quinn GP, Keough MJ (2002) Experimental design and data analysis for biologists. Cambridge Univ Press, Cambridge

Ramirez-Llodra E, Brandt A, Danovaro R et al (2010) Deep, diverse and definitely different: unique attributes of the world's largest ecosystem. Biogeosci 7:2851–2899

Schlacher T, Baco AR, Rowden AA et al (2013) Seamount benthos in a cobalt-rich crust region of the central Pacific: conservation challenges for future seabed mining. Divers Distrib. doi:10.1111/ddi.12142

Sharma R (2005) Deep-sea impact experiments and their future requirements. Mar Geores Geotechnol 23:331–338

Thiel H, Forschungsverbund Tiefsee-Umweltschutz (2001) Evaluation of the environmental consequences of polymetallic nodule mining based on the results of the TUSCH Research Association. Deep-Sea Res II 48:3433–3452

Thistle D (2003) On the utility of metazoan meiofauna for studying the soft-bottom deep sea. Vie Milieu 53:97–101

Van Dover CL (2011) Mining seafloor massive sulphides and biodiversity: what is at risk? ICES J Mar Sci 68:341–348

Yamazaki T, Sharma R (2001) Estimation of sediment properties during benthic impact experiments. Mar Geores Geotechnol 19:269–299